THE
TECHNIC __
DIVING
HANDBOOK

An updated, expanded, revised, and rewritten version of the

Ultimate Wreck-Diving Guide

The *Ultimate Wreck-Diving Guide* was the seminal book on technical diving: the primogenitor of its kind. It was written at a time when nitrox, accelerated decompression, helium mixes, rebreathers, and other emerging techniques and technologies were yet in their infancy. The publication of the original volume propelled deep diving into far deeper realms of the dark abyss.

The *Ultimate*, as it came to be called, was considered pure heresy by those who opposed progress. The purpose of the book was to introduce to the information-starved masses the small group of exploratory divers who were stretching the underwater envelope, and how they went about doing it. Call it an awareness guide.

Ignited by the author's vision, technical diving has taken off like a launched rocket, accelerating faster with each passing moment and carrying with it a payload of unknown dividends. What began as a new phase in the slow progress of underwater exploration has grown with lightninglike speed.

Evolution became revolution. Almost overnight the concepts of "high tech" and "extended range" diving entered the forefront of human consciousness.

The rite of passage is over. Technical diving has come of age.

Underwater explorers can now share the benefits of space-age spin-off hardware and developing decompression methodologies. *The Technical Diving Handbook* is a practical guide for extended range divers: a soup to nuts volume that discusses in detail every facet of this exciting and challenging underwater activity. The current edition incorporates recent innovations that were unobtainable until only a few years ago, and in some cases were nonexistent when its predecessor first saw print.

Acquaint yourself now with such new and exciting devices as the programmable nitrox wrist decompression computer, the personal computer interface, the hoseless pressure gauge, the heads-up display, the full-face mask which permits regulator switching, decompression software for laptops, the submersible electric heating pad, and more.

Also included in this handbook are tips for blending nitrox, heliair, heliox, and trimix; instructions on how to build in-water oxygen decompression stations; the procedures and the chemicals needed to clean tanks, valves, and regulators for oxygen service; and a complete chapter on how to plan and execute expedition style mixed-gas diving operations.

The previous book was the ultimate. This one goes beyond.

Lusitania
The docking telegraph lies on a slab of wooden decking which rests on the ocean bottom at 305 fsw. The letter is the first "A" in the ship's name on the high side of the bow at 270 fsw.

The
Technical
Diving
Handbook

by
Gary
Gentile

GARY GENTILE PRODUCTIONS
P.O. Box 57137
Philadelphia, PA 19111
1998

Gary Gentile Productions
P.O. Box 57137
Philadelphia, PA 19111
http://www.pilot.infi.net/~boring/gentile.html

Additional copies of this book may be purchased from the same address by sending a check or money order in the amount of $30 U.S. for each copy (postage paid).

Picture Credits

All uncredited photographs were taken by the author. Every attempt has been made to contact the photographers or artists whose work appears in this book, if known, and to ascertain their names if unknown; in some cases, copies of pictures have been in public circulation for so long that the name of the photographer or artist has been lost, or the present whereabouts are impossible to trace. Any information in this regard forwarded to the author will be appreciated. Apologies are made to those whose work must under such circumstances go unrecognized.

Disclaimer

The information presented herein is not intended to replace professional training, nor does reading this book make the reader a technical diver. The reader alone is responsible for his actions and behavior. The numbers and figures given in examples for mixes, tables, decompression profiles, and so on, are used for illustration purposes only and should not be employed in actual practice.

Acknowledgments

The author wishes to acknowledge Ted Green for the expert advice, technical assistance, and valuable insights that he provided during the preparation of this book. I would like to express my appreciation to Bill Hamilton and J. Scott Landon for generously furnishing feedback about the original volume. I also want to thank Peter and Sharon Readey for reviewing the chapter on rebreathers, and Drew Maser for proofreading the galleys.

International Standard Book Number (ISBN) 1-883056-05-5

First Edition

Printed in Hong Kong

TABLE OF CONTENTS

The Technical Diving Handbook - Table of Contents

INTRODUCTION

Bottomless cenotes, extensive cave systems, and deep water shipwrecks that were once considered inaccessible are now visited regularly. Dives to depths and durations of dives once thought of as mad are now commonplace. New tunnels and historic wrecks long lost are being discovered at an ever-quickening pace.

Technical divers are moving onward and downward.

Naturally, this progression of events did not occur overnight. It has been a hard uphill pull not only against the impediments ordained by nature but against an entrenched protocol established by those who disavow further advance.

Against nature we have science and mechanical invention, against the Neanderthal mentality we have the indomitable human spirit to succeed against all odds. Mankind rides an unstoppable treadmill of progress. Nature beware, obstructionists watch out.

Space ships were invented in the human mind when man first saw the eagle fly. Scuba diving was conceived when man first gazed in awe at the ability of fish to breathe under water and swim down into the briny depths. Between concept and actuality lies the unquenchable process of thought: making dreams come true.

In the technical diving sense, "tech" stands for both technique and technology: the acquisition of esoteric knowledge in the use of state-of-the-art equipment. Both concepts will be discussed in the following pages to provide a blend of what is presently available for the enterprising diver who is willing to challenge the next frontier.

Tech divers or "techies" primarily dive caves or wrecks, the chief difference being whether the rocky surface is overhood or underfin.

Closing the Generation Gap

A new generation of explorers has arrived in the guise of technical divers. They are going deeper, staying longer, penetrating farther, and decompressing more than ever before: all in the name of adventure and exploration.

The 130-foot depth barrier for scuba has been invalidated just as the sound barrier was in 1947. It is nothing more than a vestigial process, a mere appendix in outdated diving manuals, and accepted only by instructional agencies concerned with liability and by the U.S. Navy and NOAA due to their backward mind set and training limitations.

Any arbitrary depth restriction is a warning to the inexperienced. The meek shall inherit the Earth only to a depth of 130 feet. The technical diver owns the rest of the underwater world.

Now is an exciting time to be involved in technical diving. The field is expanding almost daily, with a prognosis for future improvements and discoveries that is both exciting and full of potential. People want to make the best use of free time in which to pursue challenges unassailable in an otherwise secure and civilized society.

Exacting researchers are delving into musty archival files and retrieving the lore of forgotten maritime tragedies. Enterprising explorers are locating long-lost shipwrecks. Audacious cave divers are attaining new depths and stretching underwater endurance to the extreme in the search for hidden passageways.

Divers everywhere are extending their physical boundaries and are examining in person new deep-water sites once regarded beyond reach.

The philosophical question such dedication begs is: Why?

"Because it is There"

What drives today's technical diver?

When asked why he wanted to climb Mt. Everest, George Leigh Mallory replied dryly, "Because it is there."

Technical divers possess a similar resolution. They descend to great depths and go great distances under water because it is part of human nature to do the undoable, to question the unquestionable, to ponder the imponderable.

Only those with a deep-seated desire for personal achievement will devote themselves to obtaining the level of excellence and expertise necessary to explore the unplumbed reaches where the challenges to be overcome are paramount, where making the right decision at the right time is vitally important, where room for error is slight.

These people are not mere sightseers, but explorers.

Change is the Way of Life

Opening your mind to the procedures explained in the following chapters will help you become a safer, more effective diver. Awareness of alternatives in both equipment and approach will lead to more comfortable and more rewarding dives. The accumulation of knowledge is an ongoing process.

Only by continuing to move can you avoid becoming mired in the mud.

Those who refuse to grow by accepting change will be left behind. You must be open to new ideas, be willing to embrace new concepts, be disposed toward testing new gear configurations that differ from your own. Otherwise lies stagnation.

There is no static system - neither in technical diving nor in any aspect of nature or human endeavor. Technical diving has been evolving throughout the years in the form of cave diving and wreck diving. And it will continue to evolve.

Evolution ends with extinction.

The Non Standard

There are those in technical diving who advocate that they have all the answers, that they have seen the light of perfection, that they have discovered the final solution, and that further progress is futile. This is quite a presumption in a line of pursuit so young.

Already these self-styled "experts" seek to impose arbitrary standards according to their own brand of gospel. They encourage blind acceptance of their teachings. And they ridicule those who see room for improvement.

These diehard stagnationists have fallen prey to inertia. Like the dinosaurs, they have isolated themselves in a specialized niche from which there is no escape. In a field that is expanding dramatically, they have already lost the ability to meet the changes and challenges that are forming on the horizon. Instead of fomenting continued development they have established a new orthodoxy. In this inflexibility lies their eventual defeat.

I fly no such standard. My position is purely iconoclastic: I don't think we even know all the questions, much less have all the answers. I repudiate all attempts to dictate rules and regulations that are capricious and arbitrary.

All too often, minimums become maximums and standards become limitations.

The only standard that I accept is the non-standard.

Call for Back-up

The sentiment stressed the most in this book is that of redundancy. Always have a back-up plan as well as back-up equipment - and sometimes a back-up for a back-up. The dark abyss is a hostile environment where demise is only a breath away.

If "redundancy" isn't your epigraph, "stupidity" might be your epitaph.

Always have back-ups.

A Note on Usage

The Technical Diving Handbook does not profess to be a text book in the accepted sense of the term. Rather, it is an operations manual about the current state of affairs in technical diving: what kinds of equipment are being used, and what techniques have been developed in response to recent technological innovations.

This volume is structured so that each succeeding chapter and section builds upon information introduced in preceding chapters and sections. But the book doesn't need to be assimilated sequentially. Depending upon the reader's knowledge and experience, he may peruse certain pages for specific instruction or may use the book solely as a reference source.

It is customary upon the first appearance of a technical term, acronym, or initialism to explain the term or to spell out the words from which the acronym or initialism were derived. I haven't always done this because this type of book is not necessarily read from beginning to end - or even read completely through at all.

Some of these shortened forms I have assumed are part of general knowledge: words like BC and DOT. This is in deference to my American readers. My readers in other countries probably won't know that DOT stands for the United States Department of Transportation. For that reason, all acronyms and initialisms are included in the glossary. If you come across a shortened form that you don't know or recognize - or can't decipher at the moment - look it up in the back of the book.

A Matter of Choice

The garnering of knowledge and experience is a ceaseless task. It begins at birth and ends at death. An individual should grow throughout life, knowing more today than he did the day before. This implies that right now he doesn't know as much as he will tomorrow. This is a philosophical truth.

Like common sense, proper decision making results from the sum of one's experiences. I cite a recent example as a moralistic story.

The day dawned warm and placid, sunny except for a shimmering haze. With eager anticipation we loaded our gear onto a staunch aluminum crew boat that had seen strong service in the oil industry in the Gulf of Mexico before being converted to a dive charter vessel.

Soon we were headed for the wreck of the *John Morgan*, a Liberty ship sunk off Virginia Beach after a collision with the tanker *Montana* during World War Two.

Aboard was a composite group of divers of various degrees of training: one woman had been diving for a year and was making her first cold water dive, I had two decades of wreck diving behind me. Equipment spanned the gamut from wetsuits to drysuits, from single tanks to doubles. The 100-fsw depth was amenable to all.

Also on the popular site was another converted crew boat and its complement of divers, a batch as mixed as our own. Both boats grappled into the wreckage without a hitch, and deployed their divers. I was among the first to hit the water.

The current was moderately strong. I pulled myself along the bow line through a slight chop noticeable only on the surface. From five feet under water I felt only the constant push of the current. The water was a muddy brown with enough suspended fine silt to reduce visibility to eight to ten feet. I reached the anchor line, then slanted downward toward the bottom.

Wreckage came into view. Instead of dropping off the anchor line I continued on until I reached the grapnel. I wanted to check it for security reasons, and to orient myself to its location with respect to the debris field. One tine maintained a firm grip on an unmoveable I-beam. The chain provided protection against chafing. The rope extended away from any sharp protrusions and metal edges. I began my exploration of the wreck.

After half an hour I returned to where I thought the anchor line should be. I missed it in the darkness and poor visibility. I conducted a sweep, re-oriented myself with a section of wreckage I had noted previously, and again passed over the supposed location of the grapnel.

Again I couldn't find it. After a third sweep of the area, and definitely recognizing the original hooking point, I accepted the fact that I was in the right spot but that the grapnel had broken free. I was now thirty-five minutes into the dive, and accumulating decompression penalty rapidly.

I prepared to make an emergency ascent. I took my decompression reel off the back of my tanks, unfurled the attached liftbag, inflated it, let it rise to the surface, and tied the quarter-inch sisal to a high spot on the wreck.

No sooner had I done this than I noticed the line coming back down: the liftbag had torn away. The next step was to tie the line directly to the wreck and unspool the sisal as I ascended toward the surface. When I reached the 20-foot stop I realized that the surface current had increased dramatically. I had to let out more line as I was pushed down like an inverted pendulum by the force of the current - a process that was continuous.

The scope kept getting longer, the tension

sharper. I was only ten minutes into a twenty-five minute hang when I saw that the underlying line was coming off black and moldy: during the period of disuse the line had begun to rot. Finally, the strain was too much for the weakened strands, and the line parted.

Now came the time for a difficult decision. Under normal circumstances, without current and with the knowledge that the boat was firmly anchored, I might have opted to surface, swim for the anchor line, and re-descend in order to complete my hang. If accomplished in short order I could avoid the dreaded bends and conduct a miss-stop decompression: not my choice of methods, but an acceptable emergency procedure.

However, I knew there was no anchor line waiting for me. I knew that the boat was adrift. I suspected that other divers might be popping up elsewhere and might require rescue. And I knew for certain that the abbreviated decompression would result in a hit that could bend me over with pain or paralyze me at a time when I needed my strength the most: either to effect my own rescue or to remain in good health until help arrived.

I opted to conduct a free-floating decompression.

Twin tanks with 100-cubic-foot capacity each, plus a 30-cubic-foot pony bottle, gave me enough air to implement this decision. For fifteen minutes I drifted with the current away from the wreck site.

Meanwhile, the situation topside was serious. At the time the anchor line broke out of the wreck there were half a dozen divers on the bottom, two on the anchor line nearing the end of their decompression, and three making their way forward along the bow line. Without a grapnel to keep the bow into the sea, the wind blew the boat broadside where it wallowed in the trough.

The captain couldn't use the propellers to straighten the boat because of the divers clinging to the lines. People and gear on deck were tossed about like billiards on a bouncing pool table.

Within the short span of forty minutes a sudden storm had blown up and stirred the sea to a froth. Wave heights grew to three to five feet, with whitecaps crashing everywhere and wind whistling through the antennas. The boat could take the pounding, but could the people and gear?

With great cooperative effort the five divers gripping the boat's flailing lines managed to climb the pitching ladder and attain the relative safety of the deck. In addition, two divers from the other boat had been swept away and were picked up by our boat. Four of our divers who reached the surface and found their own boat missing managed to catch the trail line of the other boat; they were pulled aboard.

What is but a few seconds in the telling was a tremendous amount of effort in the doing. With great difficulty the dive masters on both boats cross-referenced their lists of names in order to determine who was up and where everyone was.

Now only two divers were missing: Peter Hess and me.

Because of the turbidity of the water I couldn't see the surface from my 10-foot depth. I had no idea how bad the sea conditions had become: any up-and-down motion was masked by the current and by my constantly changing buoyancy and overcompensation as I concentrated on maintaining the proper depth. During that time I was drifting ever farther away from the only known reference point thirty miles from shore - the wreck.

I assessed my situation. I had to accept the fact that when I eventually surfaced I might not be spotted from the boat. A diver will drift with the current, but a boat is impelled along a vector that is influenced by the wind blowing against the hull and not necessarily in line with the current.

I figured that three hours was the minimum time I would have to stay afloat before any rescue could be effected: one hour before I was definitely given up for lost, thirty minutes to scramble a Coast Guard helicopter (if one was available), thirty minutes to fly to the area, and an hour running down-current search patterns. I also realized that I might have to spend the night adrift. It was not a pretty picture.

Yet, I had made a conscious decision to accept that reality rather than to chance a severe case of the bends. Someone else may have chosen to surface as soon as his emergency ascent line snapped, and to take a chance on getting picked up by the boat right away, then on getting airlifted to a recompression chamber.

That would not have been a *wrong* decision. It would have been an *alternative* decision.

And that is the crux of the matter.

Personal Responsibility

A diver's conduct is his own responsibility.

Everyone on the boat that day was there by choice. No one was forced to leave the dock, no one was thrown off the boat into the sea. Each diver knew the potential consequences of his actions, and recognized that the sea is unpredictable. And each one acted (and reacted) to the best of his ability and experience.

There are no guarantees in technical diving. The acceptance of personal accountability is of the utmost importance in all facets of life. Along with it goes the thrill of making one's own decisions in times of stress, and seeing those decisions through to wherever they may lead. There is no one else to blame.

I was lucky. I popped up in mountainous seas so high that I couldn't see above the waves. Breakers smashed against my face mask. I confronted the current, rode the wave crests, and soon caught a glimpse of the boat in the distance. I made a slow, one-handed wave that was spotted within minutes. The boat turned and zeroed in on me. I was thankful to be wearing an orange drysuit. The captain expertly drew the boat along-side, made a lee with the hull, and tossed me a line. Willing hands helped me up the ladder.

Where was Peter Hess? Blissfully unaware of the calamity topside, totally ignorant of the worry and concern we all felt for him. He was calmly conducting a remote decompression on his own ascent line - under a liftbag that was invisible to us because the current had pulled it under the dirty surface.

Like everyone else, he had had to make a decision based on the circumstances. Those with single tanks and making no-decompression dives had come to the surface and gotten aboard a boat - any boat. Peter did his decompression according to the preset rules of self-containment.

That day was an experience that none of us would forget.

As the proverb says, "Those who go to sea for pleasure would go to hell for pastime."

I might add, "Those who dive for fun in the unknown abyss would spit in the devil's eye."

Here's spit in his eye.

The choice is yours, and so are the consequences of that choice.

GEAR IN DEPTH

"Go figure" is a statement one is likely to hear from open-minded divers who wonder what insecurities motivate others to ridicule gear configurations that differ from their own. A glance at the advertisements in any dive magazine reveals a plethora of equipment brands, models, materials, and arrangements, and for a very good reason: differing end user requirements and personal preference.

What suits the needs of a particular task may not suit the needs of another. What works well for one person may not work at all for someone else. Nearly everything is a compromise. Only in an idealistic fantasy land can a vehicle exist that hugs the curves of a dirt road at ninety miles per hour while hauling ten tons of freight, and offers effortless steering with rapid acceleration. In the real world we have Ferraris, Blazers, and eighteen wheelers, and never the thrice shall merge.

Technical diving embraces a broad range of disciplines, largely in cave and wreck exploration, but also in salvage work and scientific observation. Thus there are many different types of equipment and numerous ways of rigging this equipment for the deep. This chapter seeks to bring to the readers' attention the equipment options available to technical divers today, and to suggest ways to carry it or to adapt it for purposes that the reader may have in mind.

Nothing presented here is gospel. As I stated emphatically in the introduction, personal preference should always be your watchword. Assimilate the information offered in these pages, utilize it to your best advantage, and don't be afraid to discard what doesn't quite fit or suit your tastes. Make adjustments as necessary.

Tanks

So basic an item as the scuba cylinder might seem changeless and immutable. Tanks are tanks, you're welcome to say, with evident disdain. But compressed air cylinders have come a long way since Mike Nelson threw one over his head in an episode of *Sea Hunt*. Tanks now come in a variety of sizes, shapes, construction materials, pressure ratings, and, yes, even colors. Erstwhile fad has become highly fashionable, but with good reason, as we shall see.

Not only have tanks grown in size and capacity, not only have their valves been completely redesigned, but the very manner in which cylinders are carried has been altered to suit new circumstances. Customary back-mounting still reigns supreme, but often the backplate serves additionally as the core for a more elaborate transport system to carry side-mounted tanks, and sometimes back-mounting is ignored altogether.

Tank Size

"Bigger is better" divers are wont to say, and no dictum could be more truthful when it comes to blowing abyssal bubbles. For extremely long cave penetrations and going down on excessively deep wrecks, the cylinder with the largest capacity might seem the most appropriate. The fallacy often overlooked in this sweeping generalization is that people don't all breathe the same.

If you are one of the lucky few who are genetically endowed with a large lung capacity or a highly efficient metabolic rate, small tanks may suffice. Don't let yourself be pressured into wearing tanks that you'll never use to their fullest measure, and that might bend your back like a bow.

If you are like the rest of us, think big. The standard-sized tanks for technical diving today hold between 95 and 125 cubic feet of gas. It's difficult to be more precise than that because manufacturers are marketing tanks that have a wide range of volumes. Choosing how large a tank to purchase involves taking into account some factors other than total capacity.

One weighty consideration is how heavy a tank is on your back; another is how buoyant it is in the water. This may seem like a contradiction in terms, like military intelligence or honest politician, but it's not. Two tanks of similar volume may weigh the same on land yet have vastly different buoyancy characteristics in the water; or, two tanks of similar volume may tip the scales differently. These variations are brought about largely by the material of which a tank is constructed - aluminum or steel - by the density of that material, and by the weight of the gas they contain.

Consider the following. First, while the effect of gravity is negated under water, a tank's mass never changes. You still have to push a tank through the liquid medium: the more massive a tank, the more energy is required to kick your way up to speed. Second, despite its buoyancy characteristics, you still have to carry the tanks around in order to get them filled, to store them in your house, and to transport them from your vehicle to the dive site, be it a sink hole in the middle of a forest or a boat at the end of a long crowded dock. Climbing up a steep bank in the heat of summer or ascending the narrow rungs of a ladder during a rough day at sea require strength, stamina, and balance. Bigger might be badder.

I don't want to go too deep into specifics, but a brief overview of the subject will go a long way toward increasing your understanding of the matter, and will help you to make an informed decision about which tanks are best suited for you and for the purposes you have in mind.

Remember: every system has its trade-offs.

Aluminum Tanks

Aluminum is a lightweight element which is resistant to corrosion but not inherently strong. The sidewalls of aluminum tanks must be thicker than the sidewalls of steel tanks. The result is a reduced interior diameter and a greater overall girth, and less of a disparity in total weight compared to steel than one might intuitively expect,

especially when taking comparative volumetric capacities into account. The tanks are made longer in order to yield the desired volume for the working pressure at which they are rated: 3,000 psi for 80-cfg tanks and 3,300 psi for 100-cfg tanks.

Aluminum 100-cubic-foot tanks are *extremely* long and could present a problem for people with short torsos. The valves might bang against the back of your head, or the bottoms might protrude uncomfortably below your butt, or both. Their length might make them unwieldy.

These tanks are excessively buoyant, more so in sea water, which is denser than fresh water due to salt and mineral content. In the water they behave like salvage pontoons, especially at the end of a dive when they are closer to being empty. You have to wear so much lead to hold you down that you're a hazard to the environment and a delight to your chiropractor. When I switched from aluminum tanks to steels I dropped ten pounds off my weight belt; my lower back has been thanking me ever since.

This is not to say that inherent positive tank buoyancy is necessarily bad or counterproductive. Depending upon your mission or dive plan it might even be an advantage. Keep in mind that you and the gear you wear constitute a *system*, not two disparate parts that are temporarily wedded. Neutral buoyancy of the total system is desirable and can always be achieved by the addition of lead (if you're too light) or air in your drysuit or BC (if you're too heavy). Too much of either of these offsetting expedients produces inefficiency. Too much lead in your assets curves the spine. An overinflated drysuit or BC expands your profile, thus increasing dynamic drag. The closer you are to neutral buoyancy the better off you are. Consider these scenarios.

• Carrying more lead on your belt to offset the levitation on your back results in greater negative buoyancy *of your person* should you doff your tanks submerged. This could be helpful if you wish to squeeze through a restriction in a cave or in a wreck by pushing your tanks ahead of you. Aluminum tanks hover nicely while you remain comfortably prone on the bottom due to the weight on your waist. (Read my cautionary experience with doffing steel tanks, in the following section.)

• Aluminum tanks help to maintain a horizontal profile in the normal swimming attitude: face down so you can look at the bottom.

• During decompression, when the tanks are no longer full and become positively buoyant, you float like the control car of a zeppelin with catamaran hulls.

• Because you're carrying more weight on your person, dropping your weight belt during a crisis results in instantaneous lift without resorting to air inflation.

• If you want to reduce the amount of weight on your belt, bolt a steel or lead bar to the double tank band. (This also reduces the amount of emergency buoyancy provided by a weight belt that can be released.)

The buoyancy characteristics of an aluminum tank change during a dive. This is due partially to weight and density differentials between the valve and the tank. That is, the material of the valve is heavier and more dense than the material of the tank. This disparity becomes more pronounced as the volume of gas in the tank is gradually reduced.

As gas is consumed, the tank - any tank - becomes lighter. You can perceive this most noticeably by lifting a tank before and after filling it. Compressed air in a tank weighs about one pound per twelve cubic feet. Thus a 100-cfg tank is some eight and a half pounds heavier when it is full than when it is empty. Multiply that by two for doubles.

Reducing the volume of gas in an aluminum tank shifts the center of gravity forward. An empty aluminum tank floats upside down under water because the valve acts as ballast and the empty tank as a balloon. Throughout a dive you will become increasingly top heavy with a tendency to tip over and do somersaults (I'm exaggerating).

An aluminum tank on your back further offsets your center of gravity by its placement: the lighter end of the tank being lower (that is, closer to your butt) than the compensating air in your BC or drysuit chest area.

Oxidation and Electrolysis

One advantage that aluminum has over steel is that it does not rust; it oxidizes, but the oxidation forms a protective barrier that is non-deleterious. Fresh water introduced into a tank during the filling process will not cause pitting or otherwise damage the interior surface. This means that aluminum tanks can sail through annual inspections no matter how fluid your fills throughout the year.

The introduction of sea water is a different matter. Because tanks are normally stowed upright, water tends to pool on the bottom. Salt in the water can cause pitting under the oxidation and can eventually eat through the cylinder wall.

As Ted Green notes, "aluminum tanks are much more subject to electrolysis corrosion. Many aluminum tanks begin to leak at the o-ring gland in the neck because of electrolysis that occurs between the tank and the valve. Tanks also can be fatally pitted under stainless steel hose clamps that have not been insulated with electrical tape."

Hydrostatic Testing

Like all scuba cylinders, aluminum tanks must be hydrostatically tested every five years. This test is conducted in a reinforced pressure chamber only slightly larger than the cylinder. The tank valve is removed, the tank is filled with fresh water, and a high pressure valve is screwed into the neck of the tank. Once the lid is clamped tight onto the pressure chamber, which is filled with water, the cylinder is pressurized to 5/3 (five-thirds) its rated pressure. When the cylinder walls expand, the water in the pressure chamber is forced through a tube to a gauge. As the pressure is released, the amount of contraction is noted. The overpressurization may cause some permanent overexpansion, but if the amount falls within allowable limits, the tank passes the test.

By the way, you can readily observe this expansion and contraction on a set of doubles by sticking your finger in the space between tanks both before and after a fill. Caution: don't leave your finger in the space *during* the fill, or you might have to empty the tanks in order to retrieve your finger.

A tank can fail a hydrostatic test by not contracting enough (which means that the metal's elasticity has been compromised), or it can fail catastrophically, as did one of mine. My tank flunked the way most aluminum tanks fail to make the grade: the threads in the neck gave way (possibly because they were weakened by electrolysis). Imagine if you will the explosive boom of a valve being expelled at 5,000 psi against the inside of a steel chamber. I'm glad I wasn't there when the incident occurred. Chuck Wine, who conducted the test, told me that he had to change his shorts after the event.

Usually, aluminum tanks do not fail quite so thunderously. Most of the failures that I've seen were permanently bulged by the process, each one looking like a snake that had swallowed a watermelon, because a weak point in the sidewall expanded far too much and never contracted. Some of these tanks got jammed in the pressure chamber and had to be forcefully extracted.

This discussion of hydro failures is not intended to dissuade anyone from purchasing or using aluminum tanks. Except for that one incident, all my other aluminum tanks passed hydro successfully, and passed five years later as well. There is no evidence to support the contention that aluminum tanks have a shorter life span or suffer special problems with annual inspection or hydrostatic testing. It should be noted, however, that aluminum tanks are more likely to suffer structural fatigue if they are consistently overfilled during their lifetime.

Nor do I wish to compare aluminum tanks with dinosaurs; they are not becoming extinct. If they appear to be moving away from the accepted standards of use, it would be more accurate to note that among the majority of technical divers, the emphasis in cylinder material for primary cylinders has shifted from aluminum to steel.

Size, material, pressure rating, and color for everyone.

This is due largely to the fact that the range of sizes of aluminum tanks is not as great as that of steel tanks. Aluminum tanks enjoy widespread use as accessory tanks because of their inherent buoyancy and lower cost.

Steel Tanks

Steel tanks come in two grades: high pressure and low pressure, the terms being relative to each other. By high pressure is meant 3,500 psi; by low pressure is meant 2,400 psi.

In the old days we used lower pressure steel tanks rated at 2,250 psi. These tanks are no longer being manufactured and, anyway, had a limited capacity: 71.2 cfg when pumped to 2,475 psi, a 10% overfill that was permitted by the regulating authority, the DOT, for tanks that were stamped with a "+". These tanks were called 72's, but what's .8 cfg among friends?

Even after numerous hydros many of these tanks are still around today, some since the 1970's. Because of their low pressure rating they are useful as deco bottles for carrying oxygen. This is because pumping oxygen to high pressures can be difficult or dangerous, as will be explained more fully in later chapters.

Let's discuss the two types of steel tanks that are currently available. I will refer to them by their pressure ratings rather than by the names given to them by their distributors.

High Pressure Steel Tanks

The 3,500's are manufactured by a special process to which the DOT has assigned an exemption number. They are not permitted to be overfilled. After five years, they must be hydrostatically tested, but only to 3/2 their fill pressure (5,250 psi) instead of 5/3 their fill pressure (5,833 psi) as with other steel tanks. The higher tensile strength of the steel allows the wall thickness to be thinner than the wall thickness of low pressure steel tanks.

When these tanks were first put on the market there were very few shops that could fill them to capacity. Most compressors couldn't pump that high. Happily this is no longer the case, and now the shop without high-pressure pumping capability is the exception. Be careful of being short-filled, though. Pumping proceeds more slowly at the higher pressure range, taking longer to eke out those few extra psi. An impatient tank-fill operator or one who is busy might call it a fill too early, or not wish to wait for the gas to cool. One who is inattentive might not notice the pressure rating and assume that the tank is aluminum and therefore full at 3,000 psi. Check the tank yourself before leaving the store.

If you fill your tanks from a portable compressor, high-pressure cylinders may not be the way to go. Most small compressors that are light enough to lift cannot pump air as high as 3,500 psi; or, if they do, they are more expensive and take an inordinately long time to pump those last psi. Keep this in mind when considering your purchase options. Think about buying either a high-pressure compressor or low-pressure cylinders.

A steel 120 is taller than an aluminum 100. People who are undertall, take heed.

Low Pressure Steel Tanks

The 2,400's are rated 3AA by the DOT. This is a metal designation. They are also stamped with a "+" after the date of manufacture. Although their working pressure is 2,400 psi, the "+" denotes that a 10% overfill is allowed for the first five years. It's at the overfill pressure of 2,640 psi that they achieve the capacity in cubic feet that is advertised - rounded off to a whole number instead of given in fractions.

A 2,400 is shorter, fatter, and lighter in weight than a 3,500 of equivalent volume. But note that some 2,400's are taller than aluminum 100's and some steel 3,500's. This difference may be due to the brand of manufacture, but more often is due to the total volumetric capacity of the tank. The common misconception that low-pressure tanks are heavier than high-pressure tanks doesn't take into account the fact that the low-pressure tank in question can hold more gas than the high-pressure tank used for comparison.

Some say that as a function of design a low-pressure tank can achieve a better balance on your back.

A tank that requires a low pressure to achieve rated capacity has several advantages. Short fills are practically unknown; quite the contrary, you're likely to end up with more gas than you expected. Some divers profit from this inherent irregularity by intentionally overfilling their tanks when special circumstances warrant. That long push into a cave system or that inordinately deep dive possesses a greater margin of safety when extra gas is available; or it may make the difference between success and failure of a mission.

The practice is not recommended by the manufacturer, nor is it permitted by the DOT, but the reality is that lots of people do it. The philosophical question this situation begs is one of risk assumption. Given the tank's minimum burst pressure of 6,400 psi, most people opt to flaunt the regulations, figuring, I suppose, that shortening the life span of the tank is more economical in the long run than shortening the life span of one's person. It's a matter of simple self-interest.

If you dive in remote areas, or desire the convenience of not traveling long distances to a dive shop that is open late Saturday night when you're between day trips, high-volume low-pressure cylinders may be a smart investment because they are easily filled by portable compressors.

Low volume tanks are ideal for carrying oxygen and nitrox for deco. This is because oxygen is ignitable at high pressures. The Compressed Gas Association recommends that oxygen not be compressed above 2,400 psi. Following this guideline, a 2,400-psi tank can be filled with oxygen to 100% capacity, whereas a 3,500 psi tank can be filled to only 57% capacity.

Think about this: the used car purchased from a little old lady who never pushed the needle more than 40 mph is a far better buy than the car that was raced at 90 mph by a hotrod teenager. The nonabused engine shows less wear and tear. In a similar fashion, high pressure places more stress on equipment than low pressure.

The lower working pressure of a 2,400-psi cylinder reduces the risk of catastrophic failure of o-rings and seats on valves and regulators. It puts less strain on pressure gauge hoses so they last longer and are less likely to blow out. It has even been suggested that regulators may not only require less maintenance, but may have their productive lifetimes extended.

Steel Tank Corrosion

All steel tanks are subject to pitting if water (fresh or salt) is introduced into the cylinder and if it is not detected and removed immediately. Water can be sucked into a tank that is drained completely if the valve is then opened under water. If you suspect that water has gotten into a tank, remove the valve and visually inspect the interior for signs of moisture.

Simply pouring out the water and letting the tank air-dry will not stop the interior walls from rusting, especially when the water comes from the sea. Rinse the interior with hot fresh water, upend to let drain, then dry with forced hot air (such as that from a hair dryer). If a tank is allowed to remain wet and pitting occurs, the tank must then be tumbled.

The more common form of wet injection arises from water adhering to the valve orifice, such as might occur after a wet boat ride or when tanks are left out in the rain. These tiny beads of water will then be blown into the tank if the valve isn't purged before the tank is refilled.

Tanks can also aspirate water from compressors that are not fitted with adequate filtration or whose filters are not changed with regularity. Earth's atmosphere contains micro-droplets of water in suspension, much as the propellant in an aerosol can. Air is drawn through a series of filters and water separators before reaching a compressor's pressure chamber. That's why "good" air tastes so dry.

Poorly filtered air leaves behind a watery residue which condenses like dew on grass. A few molecules of water are harmless, but an accumulation over a period of time can have detrimental effects. (A discussion of the grades of filtration can be found in the chapter on "Expedition Diving.")

Among the worst offenders in this regard are compressor systems found on boats where space and weight loads are limited (and where the humidity is always high), and those whose operators are indifferent to the damage they may be causing to their customers' equipment.

This situation is not necessarily fatal to the cylinder, although it certainly shortens its lifespan. Generally, the annual visual inspection performed on scuba cylinders will catch the condition of rusting and flaking before it damages the sidewalls beyond the point at which it can be reconditioned.

To recondition a tank, the tank is partially filled with gravel, placed on motorized rollers, and rotated until the abrasive literally "sands" the inner surface clean. Only if a diseased tank is not treated expeditiously will pitting develop too deep in the sidewalls to effectuate restoration. Steel can be sanded only so far without leading to structural fatigue. In that case the tank must be condemned.

While checkups are required on an annual basis, the prudent diver will examine his tank's interior if he suspects he got a wet fill, or if he gets air from a seedy looking shop whose employees seem complacent or unknowledgeable. Don't be afraid to look into your tanks after a multi-day boat trip. The air is always more saturated above

the ocean or a lake. And because of space and cost constraints, boat compressors may be fitted with less filtration than shop compressors.

Lest you harbor concerns, the walls of 2,400's are no more prone to pitting than their 3,500 psi cousins. Neither 2,400's nor 3,500's have been around long enough to make any predictions about their endurance with respect to hydrostatic testing. So far, I know of none that have failed to pass their first examination. But I think it is safe to assume that continuous overfilling will shorten the life of the tank.

The Gravity of Steel Tanks

Whereas aluminum tanks act as floats, steel tanks act as ballast. The latter condition is less than ideal because ballast should be placed at the lowest point of a system, such as in the keel of a sailboat. Ballast that rides too high is inclined to make a system unstable. And so it is with steel tanks.

Like a baby's first faltering steps, wearing oversized steel tanks on your back takes some getting used to. At first you might roll onto your side and find yourself going all the way over like a capsizing freighter. You have to flail and kick hard to maintain a positive attitude. Whenever I get too relaxed on deco I find myself heeling over. Sometimes I just go with it and decompress on my back. But usually I kick and struggle like an turtle on its shell and try to get back on my stomach.

To compensate for undesirable listing, try exhausting some air from your drysuit and inflating your BC. This method works especially well with backmounted BC's called "wings" because the inflated bladders straddle the tanks from below like the lateral pontoons of a raft.

There are permanent methods of counteracting top-heaviness. A block of solid core foam or a length of PVC pipe with sealed end caps can be secured to the tank bands with straps or hose clamps. Size determines the amount of buoyancy.

Put unfamiliar equipment through its paces. When I was getting used to the feel of my new steel 100's I practiced a ditch and don. (Ted Green has warned me that the expression "ditch and don" is no longer in vogue, and that by its use I am dating my certification. I bow to his wisdom but humbly submit that although "tank removal and replacement" may be more accurate and politically correct than "ditch and don," who ever heard of playing "bovine herding persons and Native Americans"?)

As I was saying before I interrupted myself, I loosened the waist strap, slipped the tanks over my head, and lay them down on a coral reef. Imagine my surprise when I immediately flipped feet up. Only a firm grip on the manifold prevented me from soaring for the surface and exploding from the water like a missile launched from a nuclear submarine.

Because I was wearing a wetsuit in shallow water I had no need for lead in order to achieve neutral buoyancy. The steel tanks provided all the weight I could handle. As soon as I removed my tanks, however, my body no longer had their weight to keep me down.

My worst fear was that someone might see my ungainly shenanigans as I struggled to get my feet down and get my tanks back where they belonged. Being in only 25 fsw my situation was precarious, but not dangerous. I just didn't think I could live with the ignominy of bobbing to the surface, swimming back to the boat, and admitting that I had lost my tanks. (With 100-foot visibility, retrieval was not a problem.) With a good bite on my mouthpiece I fought for five minutes before finally struggling back into the harness and getting the straps cinched down. Fortunately no one shot video of my vaudeville act.

The point I'm making is that you should field test any new piece of equipment before using it in actual practice, and before placing complete dependence upon it in a situation that could possibly be life-threatening. Familiarize yourself with your gear's idiosyncrasies by making a shallow water checkout, instead of wasting an expensive boat trip to discover that adjustments need to be made.

As a general rule of thumb, never test more than one untried piece of equipment at a time.

With regard to tanks and weights, the argument could be made that technical divers are not in a position to separate themselves from their tanks. They are diving as a system. Doffing tanks in a crisis is admittedly an extreme measure. And in deep water the gas bubbles in neoprene are very much compressed, so the result of doffing doubles might not be so extreme as the one described above.

Yet predicaments have occurred in which a diver became so entangled in a net or monofilament that the only way to cut himself free was to slip out of his harness to enable him to see the entanglement. Imagine what would happen if he did this only to find himself suddenly floating inverted, or skyrocketing to the surface under penalty of extreme decompression.

Composite Tanks

I don't want to leave the subject of tanks without mentioning the experimental cylinders that are composed of an aluminum core wrapped in strong manmade fibers such as Kevlar or fiber glass. At the moment, composite tanks are difficult to come by, but this is not necessarily unfortunate because their vices outweigh their virtues in most technical diving situations, although they may have sterling uses in other industries.

The greatest benefit of a composite tank is its light weight. Under water, this lack of heaviness is directly proportional to its buoyancy: the cylinder possesses so much inherent flotation that it must have lead or steel bars strapped to its sides in order for a diver to get down.

Just as negatively significant is their rated working pressure of 4,500 psi, which is made possible by the high tensile strength of the material wrapping. Although these tanks are used in European countries, few if any dive shops in the U.S. can pump air to that high a pressure.

On the other hand, a composite tank that is secured to a set of steel doubles can help to offset the negative buoyancy of the steel. This configuration can produce a rig that is lighter in air and less buoyant in the water: the best of both worlds.

Clark Pitcairn prepares to dive in a cave wearing three composite tanks secured to his doubles. (Courtesy of Cis-Lunar, photo by Bill Stone.)

A Weighty Problem

While we're on the subject of negative buoyancy I'd like to insert a few words on weight systems. We were all taught in class about the virtues of quick-release buckles that enable a diver to divest himself of his weightbelt with a snap of the fingers. This means that the weightbelt must be placed *over* the crotch strap. But as I noted a couple of sections back, in technical diving special rules apply. The accidental and untimely loss of a weightbelt may prove disastrous.

Those who dive in caves and make penetrations into shipwrecks have long accepted this fact. The fear of being pinned against a rocky ceiling or steel overhead by sudden and irreversible positive buoyancy is the rationale for installing double buckles on weightbelts.

Weighting systems have also been swept along by the tide of technological advance. Today there are belts whose lead is carried in pouches. Some pouches carry BB-sized shot, some hold lead in small blocks or pigs. All offer quick-release mechanisms with the additional advantage of incremental weight reduction in the water. As long as they have adequate safeguards against accidental release, there's no reason why they cannot be used in technical diving applications.

Cave divers have long employed a method of subtracting weight at depth. Some of the lead they need to submerge is clipped onto their harness D-rings in the form of small individual blocks to each of which a snap hook has been secured by means of cable ties. More weight is required to get down from the surface than is needed to achieve neutral buoyancy on the bottom. At depth, they unclip the weights and leave them behind while they go exploring. These "drop weights" are later retrieved upon return.

This procedure presents one obvious drawback in the open ocean: there is no guarantee that a wreck diver can relocate the drop-off point.

The system that is becoming more fashionable in technical diving is the weight harness, also known as a Miller harness. In this system the lead is secured to a belt in the conventional manner, but the belt is supported by webbing that hangs from the shoulders in the same way that a pair of pants is held up by suspenders. The total weight is borne more comfortably by the shoulders than by the hips.

The philosophical question that the weight harness poses is: how is the weight quick-released? The answer is: it isn't. In fact, the weight cannot be released at all because the tank harness is donned on top of the weight harness. Is this advisable?

Compelling arguments can be made both pro and con. If you accept the technical diving principle that tanks and weights comprise a system, and that the possibility of the need to disrupt that system is immeasurably small, your approach may be one of pragmatic acceptance. If this approach makes you feel uncomfortable, don't wear a weight harness.

It can further be argued that the large lift capacity of modern BC's makes power inflation a more viable option for effecting fast positive buoyancy than the quick-release of a weightbelt, especially in light of the fact that once a weightbelt is dropped its weight is gone forever, whereas negative buoyancy can be regained from an overinflated BC by dumping some or all of its air. Thus inflating the BC offers a measure of control that releasing the weightbelt does not. (See the section on "Back-mounted Flotation.")

Tank Pressure versus Tank Volume

"Pressure" is the force (in pounds per square inch) exerted by a gas against a cylinder wall. "Volume" is the amount of gas (in cubic feet) that a tank contains at a specified pressure.

"Fill pressure" is the *highest* pressure to which a tank may be filled with compressed gas. "Volume" in this regard is implied to mean "volumetric capacity" or "maximum volume" and is the total amount of gas that a tank can contain at its rated fill pressure. The buyer is generally more concerned with volume than with fill pressure, but there are aspects of these distinctions that you should be aware of.

The volume of a tank may sometimes be given in liters. This is a measure of the internal space of the tank. By itself, a tank's internal space measurement is not useful information for the diver who wants to know how much gas the tank will hold. To make that determination you also need to know the tank's fill pressure. By the same token, a tank's fill pressure alone is not useful information if you don't know the measure of the tank's internal space.

Do you remember this childhood riddle? "Which weighs more? A pound of feathers or a pound of lead?" In true Ciceronian fashion I will not insult your intelligence by declaring that they both weigh the same. By the same token it may seem absurd to note that a high-pressure 120-cfg cylinder and a low-pressure 120-cfg cylinder both hold 120 cubic feet of gas. But the internal space of the high-pressure cylinder is about 19 liters, while the internal space of the low-pressure cylinder is about 14 liters.

When all is said and done, "cubic feet of gas" is the commodity you need to know. Pressure ratings and internal space measurements are merely means of perception.

Here are the relevant conversion formulas:

$$liters = (cubic feet/psi) \times 411$$

$$cubic feet = (liters/411) \times psi$$

Many tanks do not yield their advertised volume unless they are overfilled. This practice is legal and is authorized by the DOT as long as the tank is stamped with a "+" after the date of manufacture. This means that for five years from that date the tank can be filled to a final pressure that is 10% higher than its maximum rated pressure.

After five years, even though the tank passes its required hydrostatic test, it is no longer permitted to be overfilled. This effectively shrinks the tank's volume - by regulatory decree rather than by natural law - by 1/11 or .09 of its original capacity. A 100-cubic-foot tank, for example, then becomes an 89-cubic foot tank.

Advertised tank volumes are sometimes "rounded off" by the distributor or the retailer, so that what you see is not exactly what you get - but it's usually pretty close. For example, an aluminum 80 in reality holds only 77 cfg at its rated fill pressure of 3,000 psi. For competitive marketing reasons, volumes are rounded up, not down.

The maximum pressure rating is precisely defined and prescribed by DOT regulation. Aluminum tanks are not permitted to be over-filled.

Pressure Gauges are not Volume Gauges

A submersible tank pressure gauge (or SPG) reads pounds per square inch, not cubic feet of gas. This makes it difficult for the user to interpret exactly how much gas remains in a tank that contains less than its rated fill pressure. If a diver owns two tanks, one that holds 100 cfg at 3,500 psi and another that holds 100 cfg at 2,640 psi, he can easily get confused by a pressure gauge readout several minutes into a dive. What is a comfortable margin in a low pressure cylinder may be an insufficient amount in a high pressure cylinder.

There is no easy way out of this dilemma. You could try to be consistent by owning identical cylinders. Or you could carry a plastic card that gives equivalent cfg per psi for the tanks you happen to be wearing. Or you could be uncommonly alert to the differences - perhaps the best course. Until someone invents a universal contents gauge, relative to the size of the tank it is gauging, pay strict attention to gas management.

Not every dive demands a tank that is completely full. One might make a short dive for a specific purpose (to pull the hook from a wreck or to retrieve a stage bottle from a cave), or one might need only a small amount of deco gas for a few short stops.

The tables on pages 24 and 25 list by volume the most popular tanks on the market and the amount of gas each contains in increments of 200 psi. This will enable the user to compare variations at a glance, and to determine in the field whether a partially filled tank contains enough gas for an express purpose. Also shown are cylinders with lesser volumes. They may appear out of place, but I have included them because these smaller brethren are used as pony bottles, stage bottles, or as side-slungs to carry argon or deco gases (which will be discussed in later sections).

These tables are presented with a disclaimer for inaccuracies which occur for a variety of reasons. For example, different gases have different compressibility factors. And the higher the pressure of a gas, the less the compressibility. There are undoubtedly enough other variables to drive a scientist mad, and even those of us who aren't as close to the edge. Unless you plan your dives so poorly that survival hangs by a breath, overlook the minor discrepancies and accept the approximations.

The formula used to calculate the volume of a tank is a linear equation based upon the measure of liquid capacity, which is provided by the manufacturer. Since the pressure exerted by a gas is directly related to temperature, assume that measurements refer to a gas held at 70° F, or room temperature.

Need I mention that when you complete your descent and notice how much your tank pressure has dropped, a portion of that reduction in pressure is due to cold water immersion and buoyancy inflation, not to anxious overbreathing? You haven't consumed as much gas as appears at first glance.

Please note that there is no advantage to leaving your tanks in the sun in order to make the needle on the pressure gauge read higher. The volume remains the same because volume is absolute. But of course you knew that.

Tank Valves and the Captured O-ring

The traditional screw yoke fitting has been orthodox since the beginning of scuba. Among recreational divers, it is still the most widely used method for securing a regulator to a tank. But the newly accepted standard for technical diving and extended exploration is the DIN fitting. DIN stands for Deutsche Industrie Norme, and is translated as German Industry Standard. The DIN fitting utilizes the concept of a "captured o-ring."

Instead of the o-ring between tank and regulator being exposed along the edges after the wingbolt is tightened, the captured o-ring is enclosed in a metal block. This makes it impossible for the o-ring to extrude under pressure. Whereas the screw yoke can be knocked ajar by crashing into the ceiling of a cave or by sliding across the deck of a wallowing boat, the captured o-ring cannot be unseated; it lies snug in its fortified home deep inside the throat of the valve. The increased reliability of such a system is obvious.

The o-ring that forms the seat is recessed in a groove at the end of the threaded regulator stem, which screws into the valve block whose internal threads cannot be seen in this picture.

The captured o-ring can leak slowly if the material is worn or torn, or if its surface is encrusted with salt, but it cannot blow out explosively. Gone are the days of ear-popping detonations and high volume hissing that sometimes followed the pressurizing of a regulator - unless you don't tighten the first stage ram sufficiently!

The DIN valve was designed for pressures above 3,000 psi, and can be used as high as 4,500 psi. (By contrast, the traditional screw and yoke is approved to 3,300 psi.) But even low pressure steel tanks take advantage of the concept because of the added safety features and the adaptability of the fitting. Of course, a DIN tank valve requires a DIN first stage on the regulator. See "DIN Regulators" for further discussion in that regard.

A DIN valve also requires a DIN fill adaptor. During the initial changeover period in the early 1990's, most dive shops were not equipped to handle DIN fills. Nowadays the reverse is true. But if you travel a lot and visit out-of-the-way places, other countries, or dive off recreational charter boats with their own compressors, you would be well advised to carry your own fill adaptor. It's cheap insurance and well worth the investment.

Captured o-rings are not infallible. As suggested above, they can leak if they are not maintained properly. Inspect the o-rings periodically for wear and tear. Remove and rinse with fresh water if you see a white coating of salt or mineral build-up. Do not smear with grease as you would for a camera or regulator o-ring; simply keep it clean.

DIN Valve Handles and Knobs

When cave divers first started using tanks equipped with DIN valves they noticed a curious phenomenon: sometimes their gas was shut off partway through a dive. It happened once to me, so I can tell you that what appears curious from behind the keyboard was a bit more exciting half a mile inside a cave. I thought the very rocks were conspiring against me. Indeed, they were!

As one worms through restrictions and tunnels with low overhead, the DIN valve handles rub against the ceiling. Each brush with the rocky roof nudges the left valve on a set of doubles a bit more closed. Eventually, the regulator is shut off completely - a cruel jest played by aquatic trolls. That's not gneiss; never take a cave for granite!

	AL 3000	AL 3000	STEEL 2400	AL 3000	AL 3000	AL 3000	STEEL 2400	STEEL 2400	STEEL 2250
	6	13	13	19	30	40	46	66	72
400	0.8	1.8	2	2.5	4	5	7	10	12
600	1.2	2.6	3	3.8	6	8	10	15	17
800	1.6	3.5	4	5.1	8	11	14	20	23
1000	2	4.4	5	6.3	10	13	17	25	29
1200	2.4	5.3	6	7.6	12	16	21	30	35
1400	2.8	6.2	7	8.9	14	19	24	35	40
1600	3.2	7.1	8	10.1	17	21	28	40	46
1800	3.6	8	9	11.4	19	24	31	45	52
2000	4	8.9	10	12.6	21	27	35	50	58
2200	4.4	9.8	11	13.9	23	29	38	55	63
2400	4.8	10.6	12	15.2	25	32	42	60	69
2600	5.2	11.5	13	16.5	27	35	45	65	75
2800	5.6	12.4	14	17.7	29	37	49	70	81
3000	6	13	15	19	31	40	52	75	86
3200	6.4	14.2	16	20.3	33	43	56	80	92
3400	6.8	15.1	17	21.5	35	45	59	85	98
3600	7.2	16	18	22.8	37	48	63	90	104
3800	7.6	16.9	19	24.1	39	51	66	95	109
4000	8	17.7	20	26.7	41	53	70	100	115

SMALL TANK VOLUMES

These two tables provide the capacities of tanks that are most commonly used in technical diving today. The values in the left column are gauge pressures in pounds per square inch. The top two rows designate the tanks whose volumes (in cubic feet) are printed in the columns underneath. Each tank is designated by its material, stamp pressure, and the advertised capacity by which it is known.

Note that these fill capacities do not necessarily represent the precise volume of the tank at its rated fill pressure.

For example, the old low-pressure steel 72 actually holds 71.2 cubic feet of air at 2,475 psi, which is 10% above the fill pressure that is stamped on the tank (2,250). Likewise, low-pressure steel tanks yield their advertised capacities at 2,640 psi, which is 10% above the stamp pressure of 2,400. Some 3,000-psi aluminum tanks are called by values that are rounded off for the sake of convenience: the 13 holds 13.3 cfg, the 30 holds 31, the 80 holds 77.4.

All numbers shown are approximate values for a variety of reasons. Tank capacities were taken from manufacturer's specifications. Calculations were made from an equation which is linear and which is based on air, and which does not take into account compressibility factors of different gases. I have also rounded off numbers for ease in reading.

Tanks used for deco gas and bottom mix I rounded off to the nearest whole number: a decimal of .49 or less was rounded down, a decimal of .51 was rounded up. This expedient results in slight irregularities between

	AL 3000	AL 3300	STEEL 2400	STEEL 2400	STEEL 2400	STEEL 2400	STEEL 2400	STEEL 3500	STEEL 3500
	80	100	85	98	112	125	131	100	120
400	10	12	13	15	17	19	20	11	14
600	15	18	19	22	25	28	30	17	21
800	21	24	26	30	34	38	40	23	27
1000	26	30	32	37	42	47	50	29	34
1200	31	36	39	45	51	57	60	34	41
1400	36	42	45	52	59	66	69	40	48
1600	41	48	52	59	68	76	79	46	55
1800	46	55	58	67	76	85	89	51	62
2000	51	61	64	74	85	95	99	57	69
2200	57	67	71	82	93	104	109	63	75
2400	62	73	77	89	102	114	119	69	82
2600	67	79	84	97	110	123	129	74	89
2800	72	85	90	104	119	133	139	80	96
3000	77	91	97	111	127	142	149	86	103
3200	83	97	103	119	136	152	159	91	110
3400	88	103	109	126	144	161	169	97	117
3600	93	109	116	134	153	170	179	103	123
3800	98	115	122	141	161	180	189	109	130
4000	103	121	129	148	170	189	198	114	137

LARGE TANK VOLUMES

adjacent increments. These disparities became pronounced for small volume bottles such as those used for argon, so I rounded off those values to the nearest tenth in order to demonstrate the change. Otherwise, the cubic-foot values for a difference of 200 psi might appear the same.

Note that two 13-cfg tanks are shown. One is an aluminum-3000, the other is a steel-2400. Each achieves its advertised capacity at a different fill pressure.

The volume of tanks other than those listed here can be calculated from the same formula I used to create these tables: rated capacity divided by stamp pressure times actual pressure equals cfg. The rated capacity is that provided by the manufacturer, the stamp pressure is found on the tank, the actual pressure is determined by means of a (submersible) pressure gauge.

For example, suppose a 40-cubic-foot aluminum tank registers 2,200 psi on a gauge. When 40 is divided by 3,000 it yields .01333; multiply that number by 2,200 and the product is 29.333 cubic feet of gas remaining in the tank.

To obtain liters, divide cubic feet by pressure and multiply by the constant 411. For example, a steel-2400, 125-cubic-foot tank that is filled to 10% above its rated stamp pressure contains 19.46 liters of gas.

Warning: theoretical tank volumes are extrapolated to 4,000 psi for comparison purposes only, and do not represent any endorsement by the author to exceed rated fill pressures.

The right valve remains open due to the counterclockwise rotation. For this reason, cave divers generally use the left valve as the port for the primary regulator. Otherwise, in case of a mechanical failure, he won't switch to his back-up only to discover that it's been shut down. Better to have the regulator in your mouth fail to deliver and to have a breathable back-up.

To help guard against this eventuality, some people replace the stock knurled knob with a low friction knob.

DIN Valve Thread Maintenance

The threads in the valve throat require some attention, particularly if much of your diving is conducted in the ocean. Spray can leave deposits of salt inside the block. Over time, the accumulation of crystals will clog the spaces between threads and exert a binding force which retards the screwing action. (We wouldn't want *that* to happen!)

Before connecting a regulator or fill hose to a tank, and after removal, crank open the valve in order to blow out any water that may have run down the threads and collected in the block. Minute droplets will otherwise pool unnoticed and get forced into the tank the next time it is filled. For best results, stick your finger into the block during the blow-out procedure. The air from the orifice will be deflected by your finger tip, will spiral around the threads, and will blow out any moisture.

Despite these precautions, valve threads might still become encrusted. The recipe for resolution is simple. Buy a small plastic squirt bottle and fill it with household vinegar. Squirt vinegar onto the threads, let stand for three or four minutes, and rinse with fresh water. Don't let the vinegar stay on too long or it will eat the chrome off the brass - five minutes will suffice. Repeat as required. Use the same procedure on the external threads of the regulator's first stage.

DIN Plugs for Preventive Maintenance

I'm a believer in valve plugs. Too many times I've heard people bemoan the Fates for damage that could have been avoided by so simple a device as a plug. Fate, in this case, was not in control.

A DIN valve is at risk of indention whenever its block is untenanted by a regulator: in your truck, on the boat, even in your basement or garage. You lay your tanks on the bed of a pickup, slide another set of doubles across the slick metal, and right into the protruding valve block. Or your tanks rip loose from the boat's bungee cord when a rogue wave happens by. Or you haul one tank past another and bang it hard on top. Accidents occur.

There are no fender benders in DIN valve collisions, only total destruction, for a dented valve cannot be pounded out or filled in with epoxy. A damaged valve cannot be repaired, it can only be replaced. Replacements are expensive.

There are rubber inserts and caps for DIN tank valves and that work well to keep out moisture and dust. But they do nothing to prevent the kind of gross damage I have just described. For that you need something more substantial. Plastic threaded valve plugs offer stopgap protection in a pinch. They screw into the valve the same as a first stage. The plastic is hard and probably strong enough to withstand all but the most violent of bumps.

The best expedient however is the plug made from chrome plated brass. This plug is solid metal and is therefore uncrushable. Not only does it offer the best protection from sharp blows, but it comes equipped with an o-ring which seals against the valve seat. If you crank the plug down tight, it effectively prevents the accidental loss of gas should the valve handle get jarred loose. This can not only save your dive, but it can save a substantial investment in helium futures.

Vehicle vibration, gear settling, equipment shifting - all can contribute toward nudging the knob and knocking open the valve enough to let inaudible amounts of gas escape. Then you show

up for the dive with a partially empty tank and your expensive helium remixed with the atmosphere.

Prevent gas loss. Cinch down your plugs.

Your gut reaction to this might be to wonder how you're going to get the plug unscrewed with 3,500 psi shoving against it. No problem. Anyone with the strength of Superman can do it; or, if you're a weakling, employ leverage from a long-handled wrench. The unthreaded end of the plug is hexagonal like the head of a bolt in order to fit between the jaws of a large adjustable wrench, standard or metric. The plug turns hard at first until you muscle it out far enough for the gas to escape past the o-ring and out the bleeder orifice. From that point a child can twist it the rest of the way by hand.

Use the vinegar treatment to keep the plug threads clean.

Valve Protection and Collision Insurance

If you thought that roll bars were installed only in racing cars and off-road vehicles, think again. Cave divers are always concerned about slamming their tank valves into the ceiling as they duck through narrow passageways, especially when riding a scooter full-speed along a bumpy corridor. For that reason they have developed methods for protecting their valves from inevitable high-speed collision. A valve or handle that is sheered off from impact can ruin your whole afternoon.

Different types of valve protectors are available. The most simplistic is a U-shaped device which clamps onto the neck of the tank and curves up over the valve. This is widely used on single tanks, but a set of doubles can be fitted with two of them, one for each tank. The thick aluminum casting is lightweight, low-profile, and unlikely to break. It leaves the hoses free and clear. I don't recommend this type for wreck diving because the open top of the U hooks nets and monofilament with remarkable abandon.

For doubles there are the roll bar and the wire cage. In each device the primary support is a tank band, to which the roll bar or wire cage is welded. The band bolts onto the tanks in the conventional manner, above the upper backplate band. When selecting the type to use, make sure that when the tanks are on your back you can reach under the roll bar or through the mesh of the cage in order to turn all your valve handles.

The roll bar is similar to its automotive counterpart. The bar is not tubular, but a wide metal sheet that rises from the outside of one tank, curves over the top of the valves, then turns down along the outside of the adjacent tank. Some people find it difficult if not impossible to bend their arms straight over their shoulders and still have the wrist motion to rotate the outer handles. The isolator valve (see below) might also be difficult to turn.

In wreck diving applications the roll bar has demonstrated at least one weakness. Ask Bart Malone, who lunged into a steel plate and got it jammed between the roll bar and tank valve; he was stuck like a moth on a pin until his buddy pulled him free. Admittedly this was a rare event. On the other hand, the top enclosure helps to keep loose line from snagging on the valve handles.

The wire cage probably has the most universal appeal. Spidery arms reach down from above the valves and grip the band with long weld

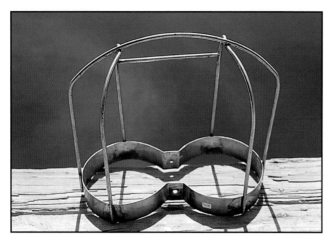

points. The open spaces allow easy access to all valve handles, including a top-mounted isolator valve. There is all-around protection from nets and monofilament. And the slender surface areas are hydrodynamically attractive even without airfoil contours. The best bet in my opinion.

Y-valves and H-valves

These are single tank valves that are used for situations in which there is no need for a supplementary gas supply, but in which a back-up delivery system is desirable. Scenarios include short, shallow penetrations, stage bottles, and deco gas. The paired ports permit the use of two regulators, each separately controlled. If there is a problem with one regulator, its valve can be shut off and the other one opened.

The Y-valve employs traditional screw yoke design. The H-valve offers lodging for DIN fitting regulators.

Manifold Destiny

Double tanks are more than two single tanks banded together. Although some people prefer to dive with independent singles, each with its own regulator and each acting as a back-up for the other - obviating the need for a pony bottle and the extra weight that a pony bottle entails - most people like to have their gas delivered continuously without having to switch regulators partway through a dive, when they might be otherwise occupied and forget to make the switch. For that reason there are manifolds.

The makers of DIN fittings did not stop at single tank valves; they designed captured o-ring manifolds for doubles as well. The basic model is equivalent to the dual port, traditional screw yoke manifold. The manifold is a straight crossover bar

that gives each port access to the air in both tanks, even if one is turned off.

It is important to understand that in this system the valve assembly on each tank does not open and close that tank; it opens and closes that port. The gas in the tanks flows freely through the manifold at all times. This means that you can breathe off one regulator throughout a dive and not worry about switching. You don't even need to put a regulator on the other port, although this practice is not recommended since a nudged knob can result in the loss of gas unbeknownst to the breather.

The beauty of the twin port manifold is that it offers complete redundancy for any first- or second-stage regulator malfunction. Should either regulator leak or free-flow, its control valve can be turned off; the other regulator is still open to both tanks and can access all the remaining gas. This doesn't mean that you should continue with your dive. The prudent person never dives without a back-up.

The Modular Valve and Manifold System

At one time single valves and manifolds were entirely different species that could not be interbred. Now a hybrid has been invented which is superior to its parental stock. This uniquely designed descendent is composed of building "blocks" far more sophisticated than those you played with as a child.

In the modular valve system, parts are not only interchangeable but additive. This allows you to upgrade from singles to doubles without expensive repurchasing, and to break down doubles into singles as needed - all in minutes and with no tools required but a wrench. The major components are basic single valves that come in left and right varieties. Opposite the knob is a threaded stem which is capped for use with a single tank. To make a set of doubles simply place two singles together, remove the caps, and join the valves with a manifold cross bar.

This system is the ultimate achievement in valve adaptability.

Isolator Valves

It is probable that sometime during your diving career you will experience a burst disc failure. Less likely but distinctly possible is slow leakage or explosive extrusion of the o-ring that seals the valve stem to the tank neck. If catastrophic gas loss occurs under water, you have only moments to reach the surface. The isolator valve cuts this problem in half.

An isolator valve does just what the name implies: it isolates the two cylinders of a set of doubles so that each is independent of the other, in effect yielding two unconnected singles. The manifold cross bar has a valve in the middle which is commonly left open but which can be closed with a few turns of the wrist the way a tank valve is closed. Instead of losing all your gas in the events described above, gas is lost only out of the tank whose valve has been compromised. You still have access to the gas in the unaffected tank: it's better than losing everything.

The modular system also offers a manifold cross bar with an isolator valve.

Systematic maintenance may obviate emergency use of the isolator system. Change valve stem o-rings annually during visual inspection, or more often if necessary. Rinse salt encrustation off valves and tank tops. And don't overpressurize your tanks - well, at least not too much or too often. Cave divers habitually overfill their tanks in order to make longer penetrations. This puts a strain on the burst disks. Let your conscience be your guide.

If you poll people who have isolator valves you'll undoubtedly hear a good deal of griping. Not horror stories, mind you, or concerns with safety in the water, but complaints about the system's one major drawback: finding the valve closed when it should be open.

People not yet familiar with isolator valve systems have a tendency to close the valve when handling tanks in group situations: a well-meaning gesture that can put you in deep water breathing from only one tank if you fail to check the valve yourself when setting up your gear. Wondering anxiously why your tank pressure gauge is plummeting so rapidly, or having to switch regulators unannounced, can be disturbing if not unnerving. The realization of truth is liable to take your breath away - but only for a moment. It's a simple matter to reach over your head and open the valve.

By the same token, dive shop operators or their employees sometimes close the isolator valve when filling doubles. This is not done intentionally, but may be due to unfamiliarity with technical equipment, or the result of years of ingrained habit, or simple lack of attention. You might not notice anything wrong until you run your gear through a pre-dive check at the dive site, and hear madly equalizing air that hisses like steam from a locomotive. If only one tank is filled, your dive is over before it even begins. Save yourself trouble by checking your isolator valve both before and after a fill.

Always check the position of your isolator valve before a dive.

Rehearsal for Handles, Messiah

Some situations call for closure before a person can get on with life in a meaningful way. Eventually there will come a time when you need to shut down a tank or close your isolator valve during a crisis. When this happens, you'd better be prepared to "handle" the trouble instinctively. Your life may depend on it.

It's important to practice manipulating valve handles, not only when you first obtain your new manifold but as a matter of routine. What else is there to do during those long decos? Make use of your time instead of just hanging around.

Some people have no trouble reaching the handles, but my arms just don't bend well enough, especially when I'm wearing thick, heavy-duty longjohns and a Thinsulate jumper under my drysuit. In order to get my hands on the knobs I have to loosen my backplate waist belt and hoist the tanks above my shoulders. (I don't secure the crotch strap.) I've left my waist strap long so it doesn't come out of the buckle during the maneuver. To recinch the belt, I simply pull the strap tight and press down on the tongue. This is easier than fishing the strap through the buckle by feel, especially when wearing mitts.

This isolator valve handle is angled slightly off vertical to enable the diver to grasp it comfortably behind his head.

All valves close clockwise as you face the handle straight on. (They do not turn left or right.) But when turning your wrist behind your neck you'll quickly discover how hard it is to figure out which way to shut down a valve: it's like looking at the back of your head through two mirrors, and trying to clip off an extra long lock. This is further complicated by the fact that because the tank valves at the sides of the manifold oppose each other, the handles must be turned in apparently opposite directions (although both still close clockwise), while the isolator valve screws down. Practice makes perfect, but there are tricks that will enable you to react faster in a bind.

When you put your regulators on your tanks, open the valves all the way then nudge them back a bit; one eighth of a turn will do. This not only inhibits the valve from sticking at the end of its travel, but it lets you know that you're turning the knob correctly: go the wrong way and it instantly snugs, go the right way and it spins about freely.

You can do the same with the isolator valve, but here you have a couple of other options to speed its shutdown. One way is to keep the valve only partially open: enough for the tanks to equalize, but not enough to put pressure on the internal assembly or to restrict the flow of gas from one tank to the other.

Another way is the ripcord technique developed by Glen Plokhoy. He wraps an abrasive string around the isolator valve knob, attaches the string to the first stage that he finds most accessible, then has only to yank the string forward in order to spin the valve shut.

Valve Handle Extender

Where there's a will there's a way, and where there's absolutely no way, along comes the will to make the way. In highfalutin technical terminology the appropriate way is via the detachable remote valve knob assembly; or, in commonplace words, the valve handle extender.

This handy device was designed to make it easy for people to crank the isolator valve shut without having to reach overhead like a contortionist. But once it came into being, creative divers found that it had other uses as well.

The basic unit consists of a flexible stainless steel cable a couple of feet long. Stiffness and black rubber wrapping make it look like a high pressure hose. On one end is an ordinary knob, on the other end is a sleeve that fits over the valve hand-wheel stem. It's equivalent to a camera cable release but is turned like a socket driver with a long spring shaft or universal joint.

The manifold isolator valve can be rotated to any position. Normally, the knob is angled up for ease in handling; in that case the cable can be run over your shoulder and secured to a harness D-ring so it hangs down conveniently on your chest. Or you can position the knob down (between the tanks) and run the cable under your arm and across your waist. Either way it puts the control knob in front of you and within easy reach. No more stretching.

The extender can also be attached to that hard-to-reach pony bottle valve handle, or, for that matter, the primary tank valve knobs. Don't get too carried away, though, or you'll have so many cables running loose that you'll look like Medusa with a head full of snakes. If you use more than one cable, try color coding the knobs.

One Pressure Gauge or Two?

This is a question that is constantly debated. One school of thought suggests that each and every tank should have its own pressure gauge. That way, if you must isolate your tanks due to catastrophic gas loss in one, you'll know how much gas remains in the other. Otherwise, if you lost the gas in the tank to which the pressure gauge was attached, you wouldn't know.

Another school disdains redundancy in this particular instance, claiming first that too many hoses and gauges get in the way, and second that, since a dive would be terminated automatically and immediately upon a malfunction that necessitated isolation, the knowledge of how little gas yet remained in the other tank couldn't help you make a more informed decision than the one you're going to make. Once you've got only one tank to breathe from, you'd better be on your way to the surface.

Potential scenarios are unlimited. It's up to the user to decide on which class to attend.

Tanks a Million

In times past, divers felt overburdened if they had to carry a pony bottle in addition to their doubles. Nowadays a technical diver may hit the water with five or six tanks of various dimensions and holding a variety of gases. These extra tanks may be used to extend bottom time, to hold mixtures enriched with oxygen for accelerated decompression, or to carry high density gas for drysuit inflation.

These cylinders are variously called stage bottles, side-slungs, sling bottles, wing tanks, saddle tanks, deco bottles, hang bottles, argon bottles, and clip-ons. The taxonomy of the moniker depends upon the way the tank is transported, the tank's intended purpose, or the way the tank is used.

"Stage bottle" has become the generic description in the vernacular of technical diving. The term originated from the cave diving practice in which a cylinder is cached or "staged" on the way into a cave and reclaimed on the way out. People say stage bottle when they mean something else, in the way they ask for Kleenex instead of a tissue. Try to be specific and avoid confusion. "Auxiliary tank" is preferred as a nonspecific description. "Sling bottle" may be used if the tank is "slung" from a harness and/or backplate.

Tank management has become an art that is still evolving in stages. The concept of diving with so many cylinders may take some getting used to, but as you add tanks to your system on a gradual basis, you'll be surprised how comfortable they are in the water and how quickly you learn to deal with them.

I don't want to put the cart before the horse,

or the tank before the pony bottle, so first we need to understand the transport system that makes it possible to carry extra cylinders with comparative ease.

Backplate and Harness

These named items go hand in hand and back to back. The backplate is part of the tank attachment assembly, the other parts being the two bands that clamp around the tank(s), and the harness that is donned like a knapsack.

Backplates are made of plastic, aluminum, or stainless steel. All three materials present strong resistance to corrosion. Some older plastic models have shown a tendency to crack or split at the strap slots, especially during traumatic entry from the high deck of a boat. Newer models made of ABS plastic are not prone to catastrophic fatigue. Metal backplates have so far proven indestructible.

Aluminum is lightweight, stainless steel is heavy; both are equally strong. Stainless steel backplates purposely add weight to the system so that lead can be removed from the belt. This not only distributes the load more evenly, but it reduces the need for an excessively heavy weight-belt. During decompression, I used to get tingles in my upper thighs whenever the lead weights pressed against the veins in my hips. Needless to say, this doesn't make for a wholesome decompression.

Some metal backplates have holes drilled along the periphery. These holes offer a way to mount argon bottles and battery packs.

There are harnesses that are sold separately. They can be worn over or under tank mounting straps that don't come with any means of gear attachment. But a better outfit is a harness whose nylon webbing not only retains the backplate, but which is rigged with D-rings to which tanks and accessories can be secured by means of snap hooks or carabiners.

Side Mount

For passing through extremely narrow restrictions, cave divers sometimes do not wear tanks on their backs but carry them by their sides in a special harness called a "side mount." In this rig, one tank is worn on each side along the thighs. Each tank has its own regulator and pressure gauge. This system gives the diver a wide, flat profile and enables him to crawl and slither through cracks.

Side mounts can be jury-rigged by wearing a conventional tank harness *without* doubles. Simply wear two side-slung bottles like auxiliaries but use them as primaries. A back-mounted BC can be added by securing it between the back-plate and a single tank adaptor from which the straps have been removed. However, the inflated BC will flutter upward like butterfly wings. Rubber retaining straps around the air bags will mitigate the flapping effect. (See "Back-mounted Flotation.")

The D-ring Circus

So basic an item as a D-ring might not seem to demand a section all its own. Yet D-rings come in various forms, sizes, and thicknesses; are secured in several ways; and serve many purposes. A short discourse on these multitudes will not be wasted.

The D-ring is a metal ring shaped like a capital D. For underwater applications it should be made of brass or stainless steel, with stainless steel being the most prevalent material construction. The most widely used size has an inside dimension of two inches along the vertical, although smaller and larger D-rings may be found. The common length is predicated upon the width of standard harness webbing, which is two inches.

Metal thickness is important. There are two popular diameters of stainless steel stock which I will designate arbitrarily as thick and thin. The stock from which thick D-rings are made is 1/4 inch; that of thin D-rings is 3/16 inch.

Thin D-rings are used as add-ons in cave diving for such purposes as securing stage bottles to a harness. These thin D-rings cannot withstand the rough treatment of ocean diving. They have a tendency to break at the weld, midway along the vertical shank, resulting in the spreading or opening of the D and consequent disengagement. Tanks have been lost as a result, or at the very least hung loose from only one D-ring for the duration of the dive.

Thick D-rings are preferred. So far they have a proven track record of ability to withstand abuse. Harnesses designed for technical diving are fitted at the factory with thick D-rings. If you have an existing harness which has thin D-rings, replace them with thick D-rings before clipping expensive or heavy equipment to them.

By means of a slide, D-rings can be mounted on webbing where they do not already exist . First run the webbing through one slot of the slide, over the vertical shank of the D-ring, then through the other slot. The slide and D-ring assembly can be loosened and repositioned quickly and easily. A minor annoyance with this system is the inclination for D-rings to fall flat against the webbing. This makes it difficult to clip on snap hooks or carabiners with one hand.

Some D-rings are welded to a slide as a unit. This way the D-rings are held away from the webbing at a preset angle, generally 45° or 90°. One handed manipulation is a whiz.

D-rings can be installed on tanks, too: both primaries and auxiliaries. To do this you need a hose clamp and a D-ring pivot bracket, both of stainless steel. If you don't use a pivot bracket, the D-ring will twist behind the hose clamp band; the torque will bend the band and eventually break it. The pivot bracket holds the D-ring perpendicular to the band and allows it to, well, to pivot.

Sometimes D-rings pivot too easily, the way they do when secured to webbing with a slide. This problem is overcome by sticking a short length of surgical tubing under the vertical shank of the D-ring and against the base of the pivot bracket. The compression of the rubber provides enough tension to keep the D-ring at an angle away from the tank, while enabling it to be pushed flat when not in use. Foam ear plugs work nearly as well as surgical tubing, although they don't last as long.

After tightening a hose clamp band, loosen it and cut off the protruding band so that, when tightened, the band ends under the bolt casing. This way you and your drysuit won't get cut on the sharp metal edge, and monofilament won't get caught under the protruding band.

For extra reliability, use double hose clamps and pivot brackets.

Hose clamps can chip the paint on tanks and cause rusting through electrolysis where the dissimilar metals touch. The former condition is a cosmetic nuisance, the latter is a bit more serious. To protect the outer wall, stretch a section of inner tube around the diameter of the tank where you want the D-rings situated, then tighten the hose clamp over it. This is especially good advice for aluminum tanks because aluminum and steel don't like close encounters.

I've gone one step further. In order to keep monofilament from snagging on the head of the hose-clamp bolt, I cover the whole assembly with another section of inner tube. I cut a wide slit in the tube so that only the D-ring is exposed. Remember to curve the corners in the slit in the shape of an oval. Right-angled cuts create stress points in the corners and sooner or later these points will tear apart.

As an alternative to hose clamps, D-rings can be installed on nylon webbing which is strapped around the tank by means of a plastic buckle or Velcro leader. Nylon will not chip paint or cause any kind of corrosion.

Now that we've got our harness and tanks rigged with D-rings, we can talk about methods for securing auxiliary tanks.

Auxiliary Tank Attachment, or
Sling Those Bottles

There's nothing wrong with being attached to your tanks. Divers usually are. Nor is any sentimentality involved. I always breathe easier when my tanks are near at hand and within easy reach. How I keep them there is the subject of this section.

For starters, let's take an auxiliary tank and rig it with D-rings as described in the previous section. One D-ring goes at the base of the tank just above the boot, the other goes at the top of the

tank just below the shoulder. To each of these D-rings a snap hook is secured by means of a device known as a quick link, made of stainless steel. The quick link is shaped like a carabiner in miniature. Instead of a gate, the quick link is opened and closed by means of a threaded sleeve.

A short auxiliary tank can be clipped right to the harness: the upper tank clip goes on the shoulder D-ring, the lower tank clip on the waist D-ring. The lower clip of a long auxiliary tank can be clipped to a D-ring at the base of the primary tank.

Lubricate the threads of the quick link, rinse with fresh water after salt water immersion, and work the nut periodically so it doesn't seize up. Spray silicone on the spring mechanism and the slider in the snap hook, too, in order to maintain glide.

Wreck divers are cautioned never to use swing gate hooks because of their affinity for monofilament. Furthermore, swing gate hooks are difficult to release when wearing thick mitts or gloves. Acceptable options are thumb actuated snap hooks, quick snap hooks (also known as butterfly hooks), and trigger hooks. Don't buy the miniatures unless you dive with your hands naked. Large hooks can be manipulated more easily when wearing mitts or gloves. And the larger the better.

Some divers use double-ended snap hooks (called dog hooks because they're used on leashes) instead of a quick link and a single ended snap hook. Let your judgment be your guide, but I don't like them because such a hook presents twice the opportunity to open accidentally during the constant jostling of decompression in rough seas, and doubles the chances of the thumb lever snagging on monofilament.

Instead of using a hose clamp and D-ring assembly below the shoulder at the top of the auxiliary tank, some divers prefer to secure a snap hook to the neck of the valve. There are several ways of doing this. One way is to install a large brass ring on the neck of the tank before screwing in the valve; in that case the valve protrusions prevent the ring from coming off. A dog clip or snap-hook-and-quick-link assembly completes the ensemble.

Another way is to knot surgical tubing around the neck of the auxiliary tank and to leave a loop extended. The rubber loop is then run through the eye of a snap hook. The surgical tubing acts as a shock absorber. It also keeps the tank close to the body so it doesn't hang loose and drag on the bottom, while allowing the tank to be stretched away for easier doffing and donning.

Note: surgical tubing rots in sea water over time. If sudden stress is placed upon rotted surgical tubing it is likely to break, usually at an inopportune time. Either replace your tubing often or use an alternative material or method of attachment.

Bungee cord is better than surgical tubing. It lasts longer and is less likely to fail catastrophically, especially if the user is observant enough to notice indications of imminent failure. The strands break one at a time instead of all at once. Usually they lose their stretch long before they break. And even then, the broken strands are held together by the outer jacket.

Yet another way is to secure a loop of narrow nylon webbing over the auxiliary tank valve. Short loops can be purchased with snap hooks sewn in place. Nylon is resistant to rotting and does not tear readily.

For ease in hauling tanks on land, for lowering into the water, and for retrieval after decompression, some divers make a sling for their deco bottles by securing a length of nylon webbing under the top and bottom hose clamps, or by running the webbing through the eyes of the snap hooks and tying the ends together. This alleviates arm fatigue since the tank is then carried on its side. You know what I mean if you've ever carried a tank by the valve; in that case, not only do your fingers ache from the grip, but the boot drags on the ground unless you bend your elbows. It's less exhausting to hold your arm straight by your side.

Dancing with Sling Bottles

As you can see, with a little know-how and the proper accouterments, keeping auxiliary tanks on your side is relatively simple. *Swimming* with auxiliary tanks will take a little getting used to. A few practice dives are recommended in order to get the D-rings placed correctly, so the bottle hangs just the way you want it. Test your rig in shallow water before jumping off into the deep dark blue. Before you know it, your awkward, lopsided cavort with sling bottles will become a graceful waltz.

You'll soon find that the most awkward part of carrying auxiliary tanks is standing up in them prior to entry, an exercise exacerbated by the rolling deck of a boat. Don't be afraid to ask for help in getting to the gunwale. But once in its element, a properly hung bottle is nearly out of the way yet easy to access.

There are obvious handicaps such as elbow room and increased drag, especially if you carry a bottle on each side. But if the bottles are slung so as to dangle horizontally, they will carve through the water like twin submarines with their snouts pointing into the path of least resistance.

Sling bottles that are *not* appropriately set cause greater drag. A sling bottle with too much leeway between points of attachment might angle lengthwise up or down. An aluminum tank will adopt a tail up (or nose down) attitude when partially empty.

It's difficult to observe the attitude of bottles clipped to your harness because they're beyond your peripheral vision. Have someone look you over in the water, then make necessary adjustments before diving again.

The more horizontal that sling bottles hang, the less drag they create.

Good form.

Another important matter to keep in mind when wearing sling bottles is balance. Just as planes fly better on two wings, two side bottles are easier to manage than one. Likewise, trim is better maintained when both side-slungs are the same kind of cylinder. Wearing a steel tank on one side and an aluminum tank on the other might adversely affect your equilibrium. Side-slungs with different buoyancy characteristics might cause a permanent list, depending upon how negative the tanks are when full or empty. You'll even notice some slight asymmetry as you breathe down one tank while the other is full.

The only way to approach perfection is to continuously fine-tune your systems.

Catch that Stage

As an experiment, I once penetrated a wreck at a depth of 230 fsw carrying two sling bottles. I was able to swim along the decks, through open hatches, into various compartments, and up and down ladders, all without difficulty. Admittedly, I did this in the *Wilkes-Barre*: an artificial reef wreck sitting upright and with little evidence of collapse, and under ideal conditions of clarity. It was fairly free of wires, cables, and loose lines that could snag.

In old wrecks, however, where natural deterioration has reached an advanced stage of breakdown and where passageways are tight and fraught with jagged, rusted bulkheads and hanging debris, a sling bottle is a detriment. Simplified and streamlined equipment is paramount.

I strongly recommend that sling bottles be left outside a wreck at the point of entrance.

Since most penetrations require a return along the same route, the tank can be retrieved on the way out. If a straight-through penetration is planned or conducted, make provision to circle back outside the hull in order to reclaim the bottle.

Some people stage their deco bottles at the bottom of the anchor line so they don't have to carry them throughout the dive, where weight and drag characteristics reduce efficiency and where grasping debris and monofilament can cause annoying hang-ups. If you decide to leave your deco bottle behind while you go roving, here are some admonitions to keep in mind.

• Make sure that you can find your way back to the grapnel. (In reality, there's no way to be certain unless you don't let it out of your sight.)

• Retain enough gas to complete decompression in case you can't relocate your bottle.

• Secure auxiliary tanks to the wreck or reef so they won't be moved by current.

• Never clip an auxiliary tank to the grapnel or chain. If the grapnel breaks free and the boat drifts away, your bottle goes with it.

Equipment Lines

If you think it's tough to stand on a rocking boat wearing doubles and a couple of sling bottles, when all you have to do next is fall over the side, try to climb up the ladder in such a rig. Superman can do it and so can Tarzan, but nonfiction characters are likely to have some difficulty. For mere mortals like us the equipment line was invented.

The equipment line in its purest form is nothing more complicated than a rope hung over the gunwale. But a few additional features will make the system more secure and a snap to use.

Start out with a non-biodegradable rope such as twisted nylon. Twenty feet will do. Splice a loop in one end so there's no chance that a hastily tied knot will come undone. Loop the rope over the rail or around a stanchion and toss the other end overboard when the coast is clear.

On the in-water extremity fasten a large snap hook or metal ring. Two snap hooks are better, spaced the same distance apart as the D-rings on your sling bottle. Best is a termination that suspends two leaders of equal length, with one snap hook fastened to each end. After decompressing, unclip one of the deco bottle's snap hooks and clip it to the *eye* of the equipment line snap hook (or to the metal ring if that was your choice of design). Unclip the sling bottle's other snap hook and clip it to the eye of the equipment line's second snap hook. Now you are free of the bottle, and the bottle is doubly secured to the equipment line. For even greater security, clip each snap hook on the equipment line to the adjacent D-ring on the sling bottle. A single snap hook can become unclipped when the line is jerked and dropped by a boat heaving hard in a turbulent sea (I've seen it happen), but quadruple clipping is unlikely to come undone.

If the in-water metal securing devices don't

have enough inherent weight to prevent the line from "sailing" in a current, secure a dive weight to the line several feet from the lower end.

Don't leave your sling bottle dangling under the boat. Pull it in right away. If the equipment line has two leaders, the weight of the sling bottle will be distributed evenly between the two clamp assemblies. This reduces the stress by half on the hose clamps, D-rings, and snap hooks.

To further diminish the jerk-and-snap action induced by savage wave heights, attach both ends of a strong bungee to the line so when the bungee is relaxed, a short open loop of rope curves out from the straight. The bungee creates spring tension. When the line is jerked suddenly, the bungee stretches and buffers the snap like a shock absorber. The loop should be measured so the rope comes up short before the bungee is stretched taut. This prevents the bungee from stretching farther than its design strength, and sundering as a consequence.

Coordinate equipment line deployment with the captain and with other divers. Don't hang ropes so close together that sling bottles can collide while dangling in the water. And be sure that before the engine is put in gear, all loose lines are pulled from the water *and* stowed where they can't wash through the scuppers. An errant line wrapped around the propeller makes for an angry captain.

Argon Bottles

Tanks that carry drysuit inflation gas are rigged differently than other auxiliary cylinders. They *can* be slung from the harness like auxiliary tanks, but for practical considerations they usually are not because the side-slung positions are already taken by deco bottles. What other locations are left?

Because of their size, argon bottles can be stashed in a variety of places that are largely out of the way: in the nooks and crannies created by the configuration of primary tanks, on the side of the backplate, atop the manifold, and elsewhere.

A bottle with a volume as small as 6 cfg - holding enough argon for a single dive - is the size of an ordinary vacuum flask. Like a flask, it can be slipped into a pocket - one that is glued to the upper leg of a drysuit; or it can be carried in a pouch that is worn on the harness waist belt; or it can be clipped to D-rings located on either side of

the waist belt buckle so it dangles horizontally in front of the lower abdomen. It can also be clipped horizontally to the D-rings on the two harness straps so it hangs over the chest or stomach area - but only if it hangs out of the way of your gauges and spare regulators.

The same size bottle can also be hose-clamped crosswise to the manifold. This can be done only after the regulators are emplaced, necessitating some last minute set-up. Care must be taken that the placement of the bottle doesn't interfere with actuation of the tank and isolator valves.

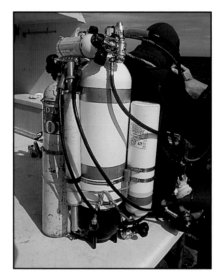

Danny Crowell has his argon bottle hose-clamped across the top of his doubles. Also of interest: the nitrox bottle mounted in the pony bottle slot, and the inverted oxygen bottle (another one on the opposite side provides balance and an alternate supply). This rig cannot be disassembled without tools, so before you think about building such a backbreaker, be sure you can climb a ladder with all the weight.

Larger bottles cannot be tucked away so unnoticeably. Those with a capacity up to about 15 cfg fit comfortably in the horizontal space behind the arm and adjacent to the diver's side and his primary tank. The backplate will offer suitable points of attachment. Rig the bottle with hose clamps and quick links (not D-rings), then add baby shackles in order to rotate the fastening angle by 90°. If the backplate is not punctured along its periphery, drill a couple of holes spaced the same distance apart as the shackles. Place the shackle pins through the holes so the bottle hangs upside down.

Argon bottles can also be worn like pony bottles, in the curved indent on the back of a set of doubles, and secured with hose clamps to a vertical bar that spans the tank bands.

An argon bottle secured to the backplate is not in arm's way.

If you already have a pony bottle mounted there, the curved indent created by the primary tank and pony bottle will suffice. Mount the argon bottle upside down so the inflator hose passes in a smooth arc under the arm.

DIN Regulators

Once upon a time there was the double hose regulator. Then along came the single hose regulator, which gradually took its place. By extrapolation we could conceivably expect the invention of a no hose regulator. But don't hold your breath.

It would be a slight overstatement to say that regulators have come a long way since the early days of scuba. Granted that today's regulators breathe easier and are more dependable than those of yesteryear, but not in the sense of quantum change. Rather, the new generation of regulators has been refined to reach performance levels that enable divers to brave the depths with greater confidence and with less breathing resistance than was available in the past.

Even the latest regulator to hit the market - made completely of titanium - is not a revolutionary concept, for its design is the same as its predecessors; only the featherweight and rust resistant material is different. At two to three times the cost of "conventional" regulators, it is not likely to make too big a dent in the sales of its competitors. More practical and economical are regulators whose critical and moving parts are made of titanium instead of stainless steel. They require less maintenance and will not get out of adjustment as quickly.

It goes without saying that DIN tank valves require the engagement of DIN regulator blocks in order to complete the DIN family. Thus the first stage of most single hose regulators has been reconfigured by their manufacturers to adapt to the captured o-ring tank valve.

Some people think that the way to save money after purchasing larger tanks with DIN valves is to use a DIN fill adaptor and a regulator with a conventional screw yoke. Do *not* do this. Not only does it defeat the primary safety feature of the captured o-ring design, but non-DIN regulators may not be constructed to handle the higher pressure to which DIN tanks may be filled. In addition, DIN fill adapters are not designed for continuous high-pressure use; they lack the reliability standards of regulator production.

The transformation from traditional screw yoke to the DIN fitting is not only one of block design. The internal workings and moving parts have been strengthened to operate at higher pressure ratings. Using a non-DIN regulator beyond its designed pressure limitation can lead to catastrophic failure at an inopportune time.

The DIN of Inequity

Just when a new standard has arrived, along comes an exception. Fortunately for technical divers DIN valves don't have as many exceptions as the English language: only two. But this bifurcation can create uncontrolled babel in foreign countries. Consider my plight when I shipped my tanks to the United Kingdom to dive the *Lusitania*, and found that I couldn't get them filled. I had to scramble madly for an adaptor so my tanks could communicate with local compressors.

There are two kinds of DIN regulators which I will designate as high-pressure and low-pressure, the comparative adjectives being relative to each other.

The regulator most widely sold in the U.S. is the high-pressure DIN 300. The number refers to the pressure rating in bars. A bar is a metric unit of pressure equal to 14.5 psi, or slightly less than one atmosphere (which is 14.7 psi). Although 300 bar equals 4,350 psi, the actual working pressure of a DIN 300 regulator is rounded up because it is designed for use on tanks whose pressure rating is 4,500 psi.

The low-pressure DIN 200 enjoys widespread use in Europe and the U.K. By strict mathematical conversion 200 bar equals approximately 2,900 psi, but again the figure has been rounded up because the regulator is designed for use on tanks with pressure ratings as high as 3,300 psi.

In both cases the double zero endings are applied as a matter of convenience for speech and composition.

The chief external difference between the two types of regulator is the first stage tank connector shaft. The shaft diameter and the thread spacing are the same in both types, but the length of the DIN 300 shaft is longer and has more threads in order to handle the higher pressure. Likewise, the throat of the DIN 300 tank valve is deeper and has more threads in order to accommodate the longer shaft.

This means that a high-pressure regulator will seat in a low pressure valve, but a low-pressure regulator won't seat in a high pressure valve because the shaft is too short. This is a safety feature to prevent a low-pressure regulator from being used on a high-pressure tank.

My tanks couldn't be filled in the U.K. without an adaptor because the valve throat was too long and the compressor's fill yoke shaft wasn't long enough to seat. Knowing this, the prudent diver will purchase DIN 300 regulators and DIN 200 tank valves. That way you'll never have this problem.

Regulator Performance

High performance regulators are essential for primary tanks and pony bottles because those regulators are employed at depth. Not so for most sling bottles, such as those carrying travel mix or deco gas. Here a run-of-the-mill regulator will suffice because it's used only in shallow water.

A high performance regulator is like a high performance engine: it delivers maximum thrust and power at the expense of frequent tune-ups. A high performance regulator is finely tuned and gets out of adjustment periodically. This doesn't mean that it's prone to failure, only that it requires regular maintenance to keep it breathing smoothly without incurring minor free-flow problems. The chief advantage of a high performance regulator is that it delivers gas at high volume with negligible resistance, regardless of depth.

On the contrary, a regulator that is produced for the masses isn't quite so quirky. It might not breathe as effortlessly at depth, or deliver the

quantity of gas, but when used within its design specifications it's likely to keep on trucking. By that definition it's unfair to think of a regulator that operates at a level that is lower than high performance as being inferior, any more than a Chevy Blazer is inferior to a Corvette. Each has a role at which it excels. I prefer to describe such regulators as "everyday" regulators, with no pejorative intended.

Rigging Primary Regulators

When a DIN block is screwed into a DIN valve a gastight seal is produced between the tank and the regulator. That might seem like the end of the story, but it's really just the beginning. The first stage can be angled at all points of the compass, the hoses can be routed in a variety of ways, and numerous modifications can be made which promote order, comfort, and flexibility in a complex arrangement of gadgets and gauges which proliferate without abandon. The possibilities are endless.

If your hoses seem "stretched" in the normal first-stage position, try skewing the block on the valve so the hoses lay more naturally, without awkward bends that create stress. Don't be afraid to switch hoses from one regulator port to another in order to achieve the desired arrangement of the devices at the end. Experiment until you get your hoses the way you want them and your devices conveniently located.

Swivels can alleviate hose tension by "bending" a device into a more suitable position. A 90-degree swivel on a drysuit inflator hose will allow it to be turned straight down under the arm instead of swooping to the side and creating a hula hoop effect. A 45-degree swivel on the second stage of a regulator will enable it to fit comfortably in the mouth without pulling against the cheek. Be creative.

Side-exhaust regulators permit great versatility in the use and placement of second stages. This is because they breathe freely in any attitude; that is, they have no upside down posture as do down-exhaust regulators. Thus a side-exhaust regulator can be routed over either shoulder, and it can be passed to someone who is facing you or who is by your left side.

I crisscross my regulators on a dual port manifold so the hoses that issue from the first stage on the right tank swing around my left side (either

over or under my arm) and vice versa. This keeps the hoses snug so there's no large loop to snare projections. The emergency shut-down procedure for a configuration like this requires constant practice in order to maintain a fail-safe routine: one must reach instinctively for the appropriate valve knob in the event of a regulator malfunction, and not accidentally shut off a working regulator while one that is free-flowing blows gas. It works for me.

A Hose by any Other Name . . .

In technical diving, hoses seem to sprout like weeds in an untended garden. The management of all these hoses requires considerable attention to architecture and constant supervision. Switching metaphors, routing hoses is like planning interstate highways, complete with flow patterns, passing lanes, and exit ramps. Everything has to work at high speed.

In technical diving, few pieces of equipment are employed straight from the box as delivered by the manufacturer. One must assume an attitude of adaptability. The key to the orchestration of hoses is modification of length. If a swivel doesn't do the job, replace an ill-fitting hose with one that is longer or shorter as the situation demands. Keep making adjustments until you are comfortable with the result.

Cave divers rig their back-up regulator on a hose that's seven feet long. This extra length and lack of constraint makes it easy to pass the mouthpiece to a friend in need without resorting to a loving embrace. Both of you have freedom of movement at a time when you need it most. The particular length was chosen because it enables two divers to swim far enough apart that they don't get in each other's way. The dependent diver can follow the lead diver from above, behind, or to the side, without getting kicked and without straying too far.

The question this begs is: what do you do with seven feet of hose when it's not in use? One way to stow such a hose is to drape it a couple of times around the neck, as if it were a scarf, and engage the second stage as your primary breathing apparatus. If your buddy requires gas, pass him the regulator that's in your mouth by uncoiling it over your head. This is the mouthpiece that a person will generally reach for - not because he is panicked but because it's the most visible and easily accessible. In this scenario you calmly go to your back-up, which should be hanging below your chin on a neck clip or lanyard.

A long hose can be looped in figure-eight fashion behind the neck. Strong rubber bands or weak surgical tubing will bind the loops against the tank tops directly under the valves. The rubber must be loose or breakable so the hose will pop free with a concerted yank. You won't have time to fuss with a hose that binds.

Another stowage location is alongside the primary tank. The hose is fed down from the first stage, looped flat against the tank to take up the excess, then routed under the arm to the chest area where the second stage is secured for instant deployment. The loops against the tank are secured by thin slices of inner tube or strong surgical tubing. The hose pops out of the loops when it's yanked.

As an extension of this idea you can connect a regulator's stages with a hose of any desired length. Low pressure hose can be purchased in coils of one hundred feet. That is how you would buy it for setting up an in-water decompression system fed from onboard storage bottles. (See "Accelerated Decompression" for a full description.) In the immediate example, picture leaving an auxiliary tank outside a wreck and then going inside on a long hose unencumbered by bulky doubles. You could fit through small openings, not incur as much risk of entanglement, and have a thick, traceable guideline to the point of entry.

As with any scheme there are disadvantages and trade-offs. You would have no way of keeping tabs on your gas supply because the pressure gauge would be on the tank. (See "Submersible Air Decompression Computers" for remote sensing types.) You risk gashing the hose on sharp metal edges. And you could not rise much above the depth of the first stage on the tank because of the pressure differential. Air delivery would not cease at once. It would become continually harder to breathe until shortness of breath would force you to halt further ascent.

A Hose of a Different Color

Color coding is helpful in many ways, from electrical wiring to piping systems, so why not use it for hose designation? I don't mean that you should scrap your basic black and buy all new hoses. There's a cheaper way of pigmenting ebony rubber coverings.

Hose protectors are available in a variety of tints and hues. In an age when color coordination has become highly fashionable, the style conscious diver can match mask, snorkel, and hose with plastic ease. (Sorry, no paisley hose.) Adding color to your hose has practical applications, too.

For example, green protectors can indicate hoses that carry oxygen, yellow can designate nitrox, orange or blue can be used for helium, and brown will signify argon. These are industry standard color designations.

Naturally, a colored protector on the first stage end of the hose is practically useless since you can't see over your shoulder (although it works okay for sling bottles). For color coding purposes you need a protector by the mouthpiece.

Better yet is colored hose wrap. This is a helical plastic coil much like the extension cord on a telephone handset. It is purchased in bulk then cut to fit a hose of any length. And it's easy to install: it can be wrapped around the hose without unscrewing the hose from either stage of the regulator.

Flavor your Gas

I gagged when I first heard about flavored mouthpieces. It sounded like another case of consumerism: a worthless accessory whose only function was to promote unnecessary sales, like wetsuits with stripes and matching designer hoods. Then I realized their true potential.

Breathing gases are as tasteless as dead baby jokes and as odorless as air. When you inhale from a regulator there's no way to know by taste or smell what kind of gas you're breathing. Differentiation is acknowledged by tank or regulator placement and by the systematic arrangement of hoses.

If you had mint-tasting mouthpieces on your primary regulators, and caramel on your ponies, you'd know right away if you stuck the wrong regulator in your mouth at the beginning of a dive or when making a switch under water. By extending the theme, you could assign different flavors to alternative gases: one for oxygen, another for nitrox, a third for helium. It certainly has potential - one that goes beyond the manufacturer's intent.

As long as we're developing means of utilizing senses that are disregarded under water, how about rose-smelling silicone masks? It beats the aroma of moldy scum that resides around my lenses.

Regulating Sling Bottles

What kind of regulator you choose to put on an auxiliary tank depends upon the tank's intended use. If the tank is filled with supplementary gas that you intend to breathe at depth, a high performance regulator is required. This holds true for pony bottles and any redundant system containing back-up gas intended for emergency use on the bottom, where high performance gas delivery might be needed.

A bottle filled with deco gas for shallow decompression, or one filled with travel gas to initiate descent, should be fitted with an everyday regulator. A suitable changeover depth is about 100 fsw.

In all cases a pressure gauge is preferred.

However, it should be noted that some people opt to do without a pressure gauge on pony bottles or other auxiliary tanks, relying instead on a pre-dive check to ascertain reserve.

You can't have hoses and devices dangling like vines from a tree or Spanish moss. They must be stowed neatly yet be instantly accessible. Rigging regulators and pressure gauges on sling bottles is a simple and painless procedure. The basic system requires nothing more than two lengths of surgical tubing cable-tied into circles that fit snug around the circumference of the designated tank. The hoses from the regulator and pressure gauge are slipped under the tubing and looped back up to the valve, where the pressure gauge can be read and the second stage pulled free. Leave a "tail" beyond the cable tie for ease in grasping.

Variations on this theme are legion. A sliced inner tube of appropriate diameter works as well as surgical tubing, but because the rubber is thin and flat it is difficult to manipulate with mitts or gloves. Since it's advantageous to restow a regulator that is no longer needed - such as after switching from one deco gas to another - make a handle or finger loop from surgical tubing or a piece of string.

Bungee cord can be bought in bulk without metal hooks. Cut it to the appropriate length as you would with surgical tubing and cable-tie the ends with a tail extended. Most other elastic materials will work as well, including thick industrial rubber bands. The latter might deteriorate more rapidly in sea water.

Because of its flat surface and the friction of rubber, inner tube slices stay in place on a tank, whereas the tubular form of bungee cord and surgical tubing cause them to roll whenever a hose is pulled out. If the top loop rolls off the shoulder of the tank it loses its grip on the hoses. This problem can be alleviated by joining the two loops with a spanner. Each loop is prevented from rolling by the drag of the other loop. The spanner can be fabricated from string or any elastic material and secured with cable ties.

The weight of the second stage may cause it to flop out of the upper loop. For this reason some people like to secure it to the neck of the tank. This can be done easily with an ordinary rubber band which you intentionally break when you pull out the regulator. A more lasting solution is a loop of elastic material (surgical tubing, bungee cord, inner tube slice) which is secured to the neck of the tank and stretched over the second stage in order to keep it firmly in place.

An excellent alternative is a mouthpiece holder which is secured to the neck of the tank by a line. This offers the added advantage of covering the mouthpiece so as to keep out mud and sand when the regulator is not in use.

A short hose on the pressure gauge obviates the need for stowage. Six-inch high-pressure hoses are readily available and are well suited for this application. I mount my first stage with the pressure ports facing down. The gauge nestles nicely against the shoulder of the tank. A swivel on the second-stage hose lets it swing around in any desired position. A second swivel on the second stage end of the hose gives infinite flexibility, and even lets me use a down-exhaust regulator on a bottle slung on my left, if I use a long hose.

Getting used to all these hoses, knowing when to reach for what, and from which tank, will keep you task loaded till you get used to the regimen. You'll fumble around a bit at first, but with enough practice sling bottles will become just another familiar feature of your diving system.

Soon you'll learn that you can wear more than one sling bottle on each side: one below the other on long and short retainers. With proper streamlining and adequate buoyancy they won't get too much in the way. You might end up looking like Robo-Diver. But who cares? As long as it works.

Inflating with Argon

So far the regulation of argon has been left up to those who need it the most. The government has not yet intervened. If you plan to dive in extremely cold water or if you want to offset the cooling effect of breathing helium mixtures, read on.

Some bottles are designed specifically for drysuit inflation gas and are assembled as a unit with a pressure reduction valve and inflator hose. Alternatively, any scuba cylinder can be used to carry argon without modification. Simply connect the first stage of an everyday regulator to the tank valve, remove extraneous hoses and plug all ports except for a low pressure port for the inflator hose, and you're ready to pump argon into your drysuit. The only item lacking is an overpressure relief valve.

In an ordinary regulator set-up the second stage will relieve excess pressure induced by first stage failure or gradual leakage, by burping tiny bubbles. Without a second stage, there's no way for excess pressure to escape. If only an inflator hose is connected to the first stage, any excess pressure leaking past the high pressure seat is forced into the hose. When the hose's containment strength is exceeded, it bursts.

This potential problem can be prevented by installing a pressure relief valve in a low pressure port. The device is the size of the end of your pinkie, and inexpensive. If the high pressure seat begins to loosen during a dive, excess pressure will be diverted through the relief valve, but you can still inflate your drysuit from the remaining gas supply.

Once when I was giving a technical diving workshop, someone in the class suggested blending oxygen with the argon in a drysuit inflation bottle, in order to produce an emergency breathing gas if all else failed. It sounds like a good idea, but don't try it.

Argon is an inert gas whose density gives it twice the heat coefficient of air. It's a great insulator against cold. Unfortunately, argon is also two and a half times as narcotic as nitrogen. This means that argox (a mixture of argon and oxygen) will make your head spin in shallow water.

I suppose if you wanted to know what it's like to dive to 250 fsw on air without incurring the risk of narcosis at depth, you could breathe argox at 100 fsw and be relatively safe - kind of like being drunk in a padded cell. You might fall down but the risk of sustaining serious injury is reduced. Nevertheless, I don't recommend it. I haven't seen argox decompression tables or chairs.

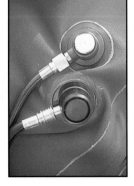

Inflation Redundancy

If the government ever learns that I've written a section on redundancy for inflation, it's likely to have me terminated "with extreme prejudice" - or hailed as the next chief of state! After all, where would we be without inflation? An economy analyst might remark, "Headed up." A technical diver might declare, "Headed down." I would retort, "Not if you plan ahead."

Since argon bottles are generally small, there's always a chance that you'll run out of gas during a dive. Or you could experience a mechanical malfunction in the argon delivery system. To avoid suit squeeze and unpleasant heaviness should such calamity occur, a back-up inflator hose is your best compensation.

I don't mean a Y-valve on your argon bottle, with redundant regulator and hose, although it is a viable alternative. The back-up technique most widely in favor is reversion to the customary inflation system: the primary tanks and regulator. Of course you won't benefit from the insulating qualities of argon, but at least you'll make it to the surface in order to complain about the cold.

You don't want to leave that secondary hose dangling out of reach. Fasten the hoses to each other with duct tape or cable ties some four inches from the end. This way you can disconnect the hose from the argon bottle while maintaining a grip on the hose from the primaries, next to be plugged in. Some people use a snorkel retainer to keep the hoses together.

Make sure to shut the valve on the empty argon bottle before disconnecting the hose from your drysuit. Otherwise water will flow up the inflator hose into the bottle. If not cleaned and dried promptly the metal might start to corrode.

If you want to go for the extra money here's an arrangement that's more convenient: install a second inflator valve in your drysuit. Now you can hit the water with both hoses connected. There's no more fumbling with plugs if you run out of argon; just press a different button. Not only does this method offer greater ease in handling, but you can adjust subjective temperature, at least to a degree. If you feel overwarm in the argon climate, squeeze the gas out of your drysuit and replenish with air. And vice versa.

And while we're on the subject,

inflation valves can be installed anywhere on a drysuit. Steve Berman installed one on his upper thigh because it was convenient to the way he carried his argon bottle. You can also relocate the exhaust valve to a place of personal convenience, then plug the hole from which the valve was removed.

It almost goes without saying that any two buoyancy systems should be inflated from different sources. For example, if your doubles are fitted with an isolation valve, the drysuit inflator hose should be connected to one tank and the BC hose to the other. That way, if you close the isolator valve and lose all the gas in one tank, the other tank can still provide gas for lift.

Electric Heating Pads

An alternative to argon as a warmth provider is a battery operated heating unit. This may not be a cure for coughs but it certainly cures the cold. In principle it works like electric socks that are worn by hunters in winter.

The rechargeable battery pack is enclosed in a watertight case that is strapped around a tank. Some users face the battery pack strap buckle inward to prevent it from coming unbuckled accidentally during the dive.

A sheathed electrical cable from the battery plugs into a bulkhead connector that must first be installed in the dive suit, through the material. An internal cable couples the bulkhead connector to the heating element, which is housed in a rectangular rubber pad that can be worn on the chest and abdomen to protect the vital organs, or on the lower back. Or both, if you have two batteries and two bulkhead connectors.

The heating pad can be worn either under a wetsuit or inside a drysuit. Since the wiring assembly contains no switch, the heating element is turned on by plugging the external cable into the bulkhead connector, and turned off by unplugging the cable. Both procedures can be effected under water with ease. Like an EO connector for camera and strobe, both male and female bulkhead connectors must be thickly coated with o-ring grease.

The unit has no thermostat or any way of regulating temperature other than by disconnecting the cable. The only complaint I've heard is that people get too warm after several minutes of continuous use in a drysuit. They have to unplug the

Joanne Surowiec is not exposing herself indecently. She is modeling the heating pad and internal wiring assembly. Notice the two wires on the inside of the bulkhead connector.

The bulkhead connector is on the midline of the drysuit above the zipper. The external cable is fed under the harness webbing so it doesn't fall out of reach when disconnected.

cable in order to cool down, then plug it back in again when they feel a chill coming on.

Gauges and the Digital Revolution

Gauges are critical components of a technical diver's paraphernalia. A diver would no more think of diving without gauges than a pilot would consider flying a plane without instruments. In this respect, about the only prerequisites that technical divers have that recreational divers have not are depth limitation and redundancy.

Because technical divers ordinarily go deep they need gauges that can withstand the same amount of pressure as the human body, and that can display meaningful information at depth. We've become a digital society and none too soon for my tastes. I remember the days before digital depth gauges when I had to rely on spring actuated gauges for determining maximum depth. Once my needle swung around so far that it passed the zero point a second time and registered 30 feet. That made it difficult to know which decompression profile to use.

Another time, Steve Gatto and Tom Packer pegged their needles so hard against the detents that the springs were permanently sprung. When they compared their gauges to my digital readout on the deco line, theirs read 20 feet too deep. In a sense, their depth gauges had become clocks - telling them that it was time to buy new and improved gauges.

Digital gauges give numbers with such precision that one is easily seduced into a false belief in their accuracy. There's a difference between precision and accuracy. A digital watch may display time in hours, minutes, seconds, and fractions - but if it isn't calibrated properly the time shown is wrong. I mention this because people tend to have an irrational reliance on digital gauges, believing, for example, that a value given to the third decimal place is more accurate or reliable than a value rounded off to a whole number. Keep that thought in mind.

There's a rule in flying: trust your instruments and not your instinct. Such a rule is valid in the air where disorientation occurs. This rule doesn't apply to diving because underwater gauges can leak or flood or otherwise go fluky. That's why redundancy is so important, and why experience teaches divers to question the accuracy of readouts that seem to be out of kilter. Learn to think for yourself.

Compasses are particularly prone to magnetic disinclination. The proximity of an iron or steel shipwreck can induce deviation from true. The needle of a compass carried next to a computer or other battery-operated device may point consistently toward the device and nowhere else. This is because all electrical fields create magnetic lines of flux which attract the seeking poles of a compass. For this reason, compasses should not be worn on a gauge panel.

When two gauges disagree, which reading do you accept? The logical answer is neither. Take the most conservative approach. An incorrect thermometer reading affects only your log book; the wrong time or depth reading can affect your decompression profile.

In technical diving, always carry at least two time pieces and two depth gauges.

A Gauge with a View, or an Illuminating Perspective

Next to reliability, the most important feature of a gauge is the clarity of its display. Some gauges display letters and numerals in dimensions that rival the small print in legal documents and prescription labels. Near-sighted divers are at a further disadvantage because their corrective lenses make the size appear smaller yet. Test the readability of a gauge before purchasing.

Numbers that appear distinct when illuminated by the sun may be difficult or impossible to distinguish under conditions of low light. This is particularly true of liquid crystal displays (LCD), which are brightened by reflected light. Numerals can be "charged" by flashing a beam of light across the face of the gauge, but the result of irradiation may not be satisfactory.

Numeric displays consisting of light-emitting diodes (LED) generate their own light; thus they are self-actuating, if you will. The semiconductor material from which the diode is made determines the color. Since the power supply is a small - perhaps minuscule - battery, the brightness is restricted and battery changes could be frequent.

Both LCD and LED gauges suffer in one respect: primary underwater lights are generally so powerful that they wash out the display (because of the eye's inability to adjust to the sudden intensity) or create glare (by reflecting off the window). Holding the light at an oblique angle doesn't always produce the desired result. You might find yourself turning the gauge like a signal mirror for brief glimpses of the display.

One solution to this problem is a reading light. This is a cigar-sized underwater light that is clipped to the mask strap or mounted atop the head. It provides just enough illumination to read gauges and slates with deco data. And because the light points wherever you look, the operation is hands-free. (Don't stare at your buddy for extended periods of time.)

Another solution to the lighting problem is internal illumination. Most digital watches have a button that actuates a light, but these buttons may be difficult to push when wearing thick mitts or gloves, and the hoped-for enlightenment may leave the viewer with eye strain. Battery life is shortened by usage.

An important feature of any gauge is a user replaceable battery - one that can be changed in the field without the use of special tools. A gauge that must be sent to the manufacturer whenever the battery needs replacing leaves you without that gauge for an indeterminate period of time.

Back-mounted Flotation

I have always rejected the idea of wearing a BC over my drysuit because the gas bag (or bladder) covers the inflator and exhaust valves and gets in the way of the hoses and disconnects. Ever since I switched to the back-mounted BC I never dive without one. Commonly called "wings", this type of BC is mounted between the tanks and the backplate. It is out of the way, is hydrodynamic, has more lift capacity than the traditional chest-mounted BC, and it doesn't squeeze the breath out of your lungs when it's fully inflated. I will abbreviate back-mounted flotation as BMF.

A high-speed power inflator can fill the bladder in seconds from the gas in your tanks, or the bladder can be inflated orally. There are at least two exhaust valves, placed so the bladder can be deflated from an upright position or by rolling over onto your side.

Some models come with a wire inside the corrugated inflator hose. This permits the exhaust valve on top to be vented by gripping the inflator assembly and stretching the hose - a convenience in any position, but considered a liability by some because of quality and design deficiencies. Detractors note that stretching the hose places undue stress on the bladder, and that failure of the valve will cause a catastrophic air loss by virtue of the valve's placement.

Wings were so-named because the bladders tend to flap when deflated. The diver soars through the water like a graceful manta ray. Some models now have a series of elastic bands around the bladder. These bands hold the deflated bladder tight against the sides, thus producing a smaller profile and reducing drag. If the bladder is inflated only partially, the elastic bands control the expansion of the bladder in order to maintain a low profile and to keep the gas distributed uniformly throughout the bladder. Without the bands, air "wallows" from side to side and continually alters lateral stability. Furthermore, the elastic bands assist in venting by helping to

Loose wings flap in the current. Rubber restraints keep the wings tight in the niche between the backplate and tanks. Note also the battery canister in its black neoprene sleeve.

squeeze the air out of the bladder when the exhaust button is pressed.

Elastic bands let a diver adjust his trim. For trim with a head-up inclination, removing bands from the top of the BMF lets that portion of the bladder expand more than the bottom. And vice versa.

A zipper in the shell allows the bladder to be removed. This enables the user to repair tears and punctures, usually in minutes, with glue and patches. If the bladder is worn out or beyond repair, it can be replaced without having to buy an entire unit.

Wings are more than a back-up for a ruptured drysuit. They handle the task of buoyancy compensation. No longer must you deal with an over-expanded drysuit that retards your forward progress, or worry about gas that might rush into the feet and flip you upside down. Air in a drysuit makes it more comfortable to wear, but by controlling buoyancy with a BMF you can avoid swelling up like the Hulk when you need that extra lift.

Once while searching for a slot in a cliff face

at the end of a shore dive I was forced to overfill my drysuit in order to keep my head above the madly crashing waves. The troughs were deep, the crests were sharp and close together. If I landed in the wrong spot I could have been dashed against the rocks. I pumped so much air into my suit that I was being choked: the neck seal squeezed my throat with a grip like that of the Boston Strangler. This condition can cause seasickness, induce vomiting, and lead to unconsciousness.

With chest-mounted BC's I always had a hard time breathing when the bladder was inflated, because the expanded bladder crushed my chest against the backplate, or added to the squeeze imparted by the tightness of a wetsuit. I also noticed a change of pitch in my voice when screaming for help because of the constriction of the crotch strap. Now I simply fill my wings to capacity and let the air in the bladder hold my head out of the water. I can float as effortlessly as a Portuguese man-of-war supported by its bell.

If I'm diving from a boat that doesn't have a ladder, a situation that requires the tanks to be doffed and tied off while I scamper over the gunwale, wings serve as a flotation device for my hardware.

Wings are also nice to have around when a drysuit inflator hose pops off, or when the inflator valve sticks open and has to be disengaged. In addition, the extra buoyancy can be used to offset the weight of heavy steel tanks, a bag full of tools taken down for a job, or to help carry large brass objects back to the anchor line.

Because of their flat design, wings can be stacked like cards in a poker deck - although I've never seen anyone wear more than two at a time. This system works as well for divers as it does for dragonflies. Two BMF's provide redundancy for wetsuit diving. However, two BMF's do not have double the lift capacity of a single BMF: due to space limitations both cannot be fully inflated at the same time.

One model goes a step further by installing dual bladders within a single shell. Each bladder has its own inflator and exhaust valves, with one bladder acting as a back-up for the other. There is not enough space inside the shell to inflate both bladders simultaneously, but this fully redundant system obviates the need to fly with two pairs of wings.

Never use the same tank for both inflation systems. Whether you wear a drysuit and one BMF, or a wetsuit and two BMF's, always connect each inflator hose to a different cylinder. That way if one tank runs dry or if you have to shut it down during a dive, you'll still be able to obtain positive buoyancy.

The sole complaint commonly made against wings is that they do not guarantee that an unconscious diver will automatically float face up: there's about a 50-50 chance. This condition is inherent in the design. It matters not whether a cave diver hits the ceiling with his nose or with his back. But in the open ocean it is definitely advantageous to be looking at the sky.

Wings are a tool designed for flotation and buoyancy compensation, not as a lifesaving device. Whether you decide to wear one is your choice.

Full-face Mask

There hasn't been a revolution in the development of masks, but one style that is coming more into vogue these days is the full-face mask. Once used only in the realms of commercial and military diving, technical divers have discovered the potential benefits of this face saving device.

As with any other piece of equipment there are advantages, disadvantages, and special-use applications, none of which has anything to do with vision. Let's take these conditions in reverse order.

Full-face masks are worn primarily when diving in contaminated water and in conjunction with wireless communication units. The former is a condition that is not often encountered in technical diving, and the latter is given full coverage in the following chapter.

The first objection you're likely to hear concerning full-face masks is that there's no way to change regulators during a dive, thus negating an essential emergency convention that technical divers rely upon. While the statement is not strictly true, switching regulators is at the very least a major inconvenience.

Jim Baden demonstrated to me how he switched to surface-supplied oxygen regulators during the decompression phase of a dive. He inhaled deeply, closed his eyes, removed the full-face mask, inserted the replacement mouthpiece, pinched his nose, purged and inhaled, then donned and cleared his spare mask. The whole procedure took less than five seconds. If you can remember far enough back, you undoubtedly performed a similar exercise during your basic scuba course.

Why does Baden pinch his nose when we are trained to breathe through our mouths with our masks off? First of all, when cold water hits a diver's face it can trigger the gagging response that makes it difficult to inhale while trying to prevent water from coming in through the nose. Pinching the nostrils eliminates the problem while allowing the diver to concentrate on breathing.

Secondly, habits change with continued use of a full-face mask, when you can breathe naturally through your nose instead of having to breathe through your mouth: an expedient that also helps prevent your mouth from going dry.

If you're like me, you did mask removals in the pool as part of your course work and never did it again. Perhaps it's another one of those exercises you should practice during those long, boring hangs, until the drill becomes second nature. Or perhaps you don't want to place yourself at risk with a ceiling over your head, in case you drop your mask; practice when your decompression is over.

Although the spider harness grips the head like the fetal creature in *Alien*, and can't be knocked off accidentally by bobbing anchor lines or by crashing into partitions and cave walls, it can be removed quite readily by pulling upward and outward from under the chin. Make sure you keep your eyes closed if you wear contacts. You can let the mask go: you can't lose it because it's attached to your primary regulator. So you see, you don't really lose your redundant gas supply, you just access it differently.

About the only other factor you need to be aware of is the possibility of carbon dioxide (CO_2) build-up. This could occur in older models because of dead spaces in the mask where gas circulation was poor. Nowadays masks are designed to minimize this condition: incoming gas continually flushes out residual CO_2.

In some models a flexible rubber cup (commonly referred to as an oral/nasal) seals around the mouth and nose. Inhaled gas is forced up along the face plate into the large visual cavity, then sucked in through internal ports into the mouth and nose. This directional flow has the additional effect of automatically defogging the lens.

Full-face masks are warmer than traditional masks because facial skin is not exposed to the water, and because the gas space retains heat.

Constant positive pressure prevents flooding. The full-face mask is also self-purging. As you pull it off your face the mask immediately forces out all water. Should you knock the mask ajar, or decide to put it back on after removal, it will dewater in an instant. Naturally, all these procedures require some proficiency. You'll pick it up just like you learned to clear your mask that first day in the pool.

One distinct advantage to wearing a full-face mask is that an unconscious diver will continue to breathe normally inside it.

The Manifold Block

Instead of switching regulators, another way of accessing back-up or alternative gases when wearing a full-face mask is through a manifold block. In this system the regulator that is built into the mask is fed from a supply hose which is connected to the output port of a metal block. The generic block is configured like an X and has three additional ports for input: one for the hose from the primary tanks, and two that can be used for redundant gas supply or deco gas.

The manifold block derives from the surface-supplied commercial diving rig which permits emergency gas from a bail-out bottle to be fed to the helmet or full-face mask in case the umbilical hose ceases to deliver gas. The technical diving model features quick-disconnect sockets and hose fittings that work with the same facility as a BC inflator hose.

Proponents who use the system rave about the elegance of switching gases with a quarter-turn flick of the valve handle - the regulator never leaves your mouth. Alternative gases are regulated by their own first-stage, but they all use the same second stage in the mask. For in-water oxygen decompression, the dangling hose can be fitted with a quick-disconnect socket instead of a second stage and mouthpiece. Simply plug in the surface-supply hose, twist the valve, and continue breathing.

The three input ports in no way limit the number of gas feeds that a diver may carry. It is a simple matter to disconnect any input hose from the socket on the block and connect another hose in its place.

Detractors point out that on the bottom it is deceptively easy to make a mistake and switch to the wrong gas - and breathing a deco gas at depth may prove fatal due to the high partial pressure of oxygen. While proponents don't deny that this can occur, they also sagely advise that, like any other piece of equipment, methodical training and practice are required in order to instill instinctive handling of the system.

Detractors also note that the system employs only one second-stage, thus violating the redundancy rule: if the second-stage fails no back-up is available. This is only partially true. A quick disconnect plug can be attached to a second-stage hose socket, providing not only a back-up second-stage but permitting a buddy wearing standard equipment to access the gas of a diver using a manifold block.

SuperMask

The SuperMask is the latest development in full-face masks. Its modular design permits gas switches without a manifold block and without complete removal of the mask. Co-designer Connie Morgan describes it succinctly:

"The Morgan SuperMask has a full face seal that runs across the forehead, down each side of the face and around the chin. The inside of the mask has a seal that runs across the face just under the nose. This makes the upper part of the mask (eye space) completely isolated from the lower part (mouth space). The eye space

functions identically to a typical scuba mask (half mask).

"The lower part of the mask (mouth pod) is flexible rubber that is mounted on a solid frame. Any demand regulator or rebreather T-Bit can be used. The choice to use or not to us the mouthpiece is up to the diver. The flexible rubber allows the diver to push in the mouthpiece and hold it there any time.

"The mouth pod unlocks with two simple catches that can be easily operated with gloved hands. A hinge at the bottom of the pod allows the mouth pod to completely open without detaching. When the mouth pod is open, the diver can breath from any standard regulator that has a mouthpiece. It can be his/hers own octopus or anyone else's. When the mouth pod is detached, the regulator can be used to buddy breathe with another diver by using the mouthpiece.

"The mouth pods are interchangeable. This allows the diver to use other breathing gas supplies without losing the full face mask configuration. Pure O_2 can be used for decompression by switching to a separate mouth pod attached to an O_2 cleaned supply system and regulator. The only part of the face exposed during an in-water gas switch is a small area around the mouth."

"Voice communications, isolation from contamination or cold water, increased diver safety and all the benefits of a full-face mask are present without compromise."

There's nothing more I can add.

Spare Mask

If this were a novel, the astute reader would undoubtedly balk at the contrived plot device in which Jim Baden magically produced a spare mask to don after removing his full-face mask, because I didn't foreshadow the

event. In reality, a spare mask is a common accessory under certain circumstances. Cave divers generally carry one, and so would a technical diver wearing a full-face mask.

A spare mask can be carried in a variety of ways. It can be strapped to a light or around an arm or leg. It can be concealed in a pouch or pocket that is glued to the dive suit leg, worn on the harness webbing, or clipped to any convenient D-ring.

The most ingenious place to keep a spare mask is around the neck facing the back, where it sits snug against the nape. Only a quick twist is required to position the mask under your chin. Then when you yank off the full-face mask, the spare can be pulled up in place without further ado. With this stowage technique the mask cannot be lost or dropped.

Chuck Schmidt carries his spare mask on his nape when diving with a full-face mask. The wire protruding from the mouthpiece is for the communication unit.

In a different scenario, a diver wearing a traditional mask might carry a spare in case the primary mask is lost or broken. A mask can be lost if the strap breaks, if it is dislodged by snagging the anchor line or bumping into another diver, or if a swift current snatches it off your face. The faceplate can be shattered by smashing the lens against a rock or bulkhead. In any of these or other eventualities, you simply spit out your regulator in order to get the mask past your mouth, reinsert the regulator, clear the mask, and off you go.

Since a spare mask is intended to be donned under water, it should be of low volume.

. . . And Then There Was Light

The amount of illumination available for underwater service rivals the dawn's early light, and approaches the brightness of the sun at high noon. (Is there a low noon?) The market is expanding with an array of powerful dive lights whose high-intensity bulbs are made from rare and exotic elements and which are powered by self-contained voltaic cells or from large external battery packs.

Choosing an appropriate light is likely to leave one in a quandary. There are lights for every purpose and for every wallet or pocketbook. Unlike most commercial products, dive lights do not feature bells and whistles to distinguish one from another. The differences are purely practical. Here are some guidelines to help light up your life.

There are two basic types of dive light: one that can be held in the hand as an independent unit, and one whose lamp assembly and battery canister are separate units connected by a waterproof electrical cord. The latter I will refer to as a "canister light." Hand-held lights predominate in wreck diving while canister lights prevail in cave diving.

When selecting a light, the most important thought to keep in mind is the kind of diving you intend to do. In that regard, ask yourself which features you need in a light and which ones will be impediments. Then compare the lights on the market with respect to physical dimensions, weight, output, and burn time; and whether you prefer throwaway batteries or rechargeable. Let's discuss these options one at a time.

• Physical dimensions: Size is usually not significant with respect to hand-held lights, but canister lights can add drag to an already overloaded diver who might have to swim against strong current and through pounding waves, and who might already be overburdened with sling bottles and argon.

• Weight: Battery canisters cannot be detached readily underwater, so gravity will work against you if you have to climb a boat ladder in heaving seas.

• Output: The more light you have, the better you can see. Buy as many lumens as you can afford, in a package whose weight you're able to carry.

• Burn time: More important for cave divers who make long penetrations. Wreck dives are usually shorter due to the uncertainties of decompressing in the open ocean, and wreck penetrations are less extensive. However, in addition to "dive" time wreck divers should consider "trip" time: that is, the duration of trips on boats which remain at sea overnight - or for several days - if no electricity is available for recharging batteries.

• Throwaway batteries: Inexpensive, obtainable everywhere from drug stores to supermarkets, and discharge more slowly than rechargeable batteries (meaning that they provide light longer than rechargeable batteries before the latter are drained and need to be recharged.) Unfortunately, throwaway batteries can't power high-intensity bulbs so the light is always dim by comparison.

• Rechargeable batteries: High-priced, available only from dive shops, short duration, and incredibly powerful; a must-have for the conditions that technical divers encounter. The problem, of course, is that they have to be recharged, sometimes between dives if a repetitive dive is planned. Some types can be recharged in three or four hours, others require fifteen hours or more for a full recharge. Generally, the larger the battery, the more time required for charging.

• Back-up batteries: Hand-held batteries cost nearly as much as the light, whereas canister batteries are cheap enough to buy a spare or two.

Make sure that any light you take to depth is rated for the pressure.

The Light Switch
As with any flashlight, the weak link in a dive light and the part most liable to fail (other than the bulb) is the switch assembly. The good news is that dive lights are built to specifications far more rigid than their air-born counterparts and are tougher and far more reliable. This is not to say that switches don't go bad, only that it probably won't happen till the light is several years old.

No switch is infallible, whether it be a magnetic switch or a mechanical switch. Mechanical switching requires that a hole penetrate the casing for the insertion of a stem or lever. The hole is sealed either by o-rings or a rubber cap. In either case, the waterproofing mechanism of the switch assembly is more prone to failure than the switch itself. A failure of this nature leads to leakage or massive flooding.

If you detect signs of moisture in the casing or canister, and the primary o-ring appears unworn and properly placed, immediately suspect the internal o-ring in the switch assembly or micro-cracks in the rubber cap. A light with a malfunctioning switch should be returned to the manufacturer for repair. The cost of the light warrants the expense.

Some models do away with switching altogether by closing the circuit with a twist. When the lens holder is rotated clockwise, it advances down a set of threads and brings the bulb in contact with the battery. Counterclockwise rotation separates the two, usually by means of a spring. Don't unscrew the lens too far or it will come right off and flood the casing. I suppose you could call this a rotary switch, although there actually isn't any switch at all, just a rotary motion.

The Switch Lock
If an ordinary flashlight is turned on accidentally it's a minor inconvenience. But a discharged dive light can result in a missed dive or at least a change in plans and a mad scramble to borrow a spare. And that's not the worst of it.

The bulbs of technical lights shine with such intensity and heat that they cannot be operated in air without the cooling effect of water. You can switch a light on for a second or two to make sure it's working, but if you leave it on any longer it will melt the plastic lens - and it might destroy the battery. If the light is packed with your gear, it will then set fire to whatever is in its path - like a ray gun or laser beam in Darth Vader's hands.

Recognizing these hazards, some manufacturer's have installed a locking mechanism on the switch. I say "some" because not all models have locks. And one manufacturer whose light does have a lock holds the mechanism in such low regard that it recommends - in writing - that the battery be turned around inside the casing to prevent accidental actuation.

I suggest that you follow the recommendation. I also suggest that you get in the habit of checking your lights before entering the water, or you might find yourself on the bottom with a light that won't come on because the battery is installed backward. Perhaps some day this overt and acknowledged design flaw will be re-engineered.

Some canister lights have a switch guard to protect the toggle from being bumped on or off. They do not have a lock.

The light of perfection is a matter of lock.

Hand-held Lights

"Heavy light" may seem like an oxymoron, but it's true nevertheless that some hand-held illuminators weigh a considerable amount because of the size and density of the battery. Not that this makes them unwieldy to handle, but I've found that all that poundage hanging from my wrist by a lanyard can be exhausting on my arm.

Dive lights are easier to carry clipped to a harness D-ring. Secure a lanyard to the handle and attach it to a snap hook or carabiner as you would with a sling bottle or other piece of equipment.

Two lanyards of unequal length add versatility to the rig. Snap both to the D-ring or to different D-rings. The short lanyard keeps the light close to your body during travel. To use the light, unclip the short lanyard and stretch the light to arm's length with only the long lanyard clipped. This way if you drop the light or let go of it, you won't lose it.

Canister Lights

Canister lights are an encumbrance to carry at best, but they provide so much illumination and endure so long on a single charge that in appropriate circumstances they are well worth their weight. Most of the weight is concentrated in the battery, which is housed in a waterproof canister. This canister can be secured on your person in several ways.

It can be slipped into a pouch which has a belt loop through which the harness waist strap is passed, although you might find that this gets in the way of your elbow. The pouch can also be secured by means of clips or quick links to holes in the backplate, in the same manner in which you would secure an argon bottle. This places the canister farther back and more out of the way. If you carry an argon bottle on one side, a battery canister on the other side will help to maintain balance.

The canister can also be hose-clamped to your tanks like a pony bottle, or in the notch created by the pony bottle mounted to your primaries. In this position it might offset the weight of, say, a decompression reel.

A favorite hiding place that cave divers like is under the tank bottoms - butt mounted, as it is called. The placement seems awkward, and you might not like the weight resting on or bouncing against your butt. But the method takes advan-

tage of an un-utilized location. If you don't mind the constant spanking - or enjoy it - here's how to rig the system.

Secure two D-rings near opposite ends of the canister. Employ hose clamps and swivels as you would with a sling bottle. If you already have D-rings secured near the bottoms of your primary tanks, for use with sling bottles, you can get double duty out of them. Otherwise, add them. Now cut a pair of lanyards or stout bungee cords to a length which, after snap hooks are added, will allow the canister to hang below the tank bottoms. The idea is not to have the canister snugged tight to the tank bottoms, but to have enough slack so it can be pushed out of the way when you sit.

For donning purposes, the canister is clipped behind the tanks while the tanks are standing upright. After slipping your arms into the harness, stand up, and when the tanks come off the dressing platform the canister will swing into place. Care must be taken when sitting down because you don't want the weight of the tanks resting on the canister. Reach behind and sweep the canister out of the way before sitting.

The canister can be padded as an extra precaution. If it doesn't come with its own pouch, make one out of neoprene. Also, feel free to experiment with alternative paraphernalia and means of securement.

A valid argument can be made against butt

mounting a battery canister for ocean diving because the dressing platform is in constant motion and a diver's stance is never stable on a rocking boat whose deck is wet. After clambering over the top of a ladder and duck-walking across a pitching deck, divers are wont to crash into a seated position with the aplomb of a ball player sliding across home plate trying to beat the throw.

So someone came up with a bright idea of mounting a roll bar on the bottom of the tanks instead of - or in addition to - a roll bar on top. Now the canister is protected in its own private enclosure.

The lamp assembly can be clipped to any convenient D-ring, snapped into a holder that is mounted on a helmet, or fitted onto a bracket that slides over the hand. Wreck divers beware of long looping electrical cords that might snag or get sliced on wreckage. Maintain a tight profile.

Manufacturers of canister lights offer options that you should consider before purchasing. Bulbs may be available with different outputs: a brighter bulb will have a shorter burn time, whereas a bulb of less intensity will last longer before draining the battery. Different battery sizes may be offered. Generally a larger battery does not supply higher voltage for a brighter light, but has more capacity and will power the bulb longer.

Bulb assemblies exist in two basic types: exposed bulb and sealed bulb. A large reflector increases the amount of light thrown forward from an exposed bulb, although the diameter of the reflector may prove disadvantageous for most wreck diving situations. The sealed beam assembly is smaller and more rugged; light is focused through the lens in order to increase efficiency.

A canister light's electrical cord can cause problems. Not only might it get in the way or snag on jagged protrusions, as noted above, but its waterproof jacket can develop leaks, with consequences that are dire. Water then not only shorts out the circuit, but the pressure may force it through the internal raceway into the bulb assembly or battery canister, flooding everything. Before deciding on which model to buy, ask the dealer or manufacturer if the bulb assembly and battery canisters are sealed from the electrical cord so an event of this nature cannot occur.

The spiral hose sleeve used on regulator hoses can be wrapped around the electrical cord for abrasion protection.

"...Nor Any Drop to Drink"

Samuel Taylor Coleridge said it like it was in his famous poem, "Rime of the Ancient Mariner." (Coleridge was a better poet than he was a speller; he misspelled "rhyme." He didn't mean to imply that the Ancient Mariner was covered with frost.) The section quote begins with "Water, water, every where . . ." and refers to a crew of shipwrecked sailors who are slowly dying of thirst.

Health conscious individuals who speak with bourgeois affectation may enjoy exploiting the language with quasi-medical pretentiousness. Thus they will say "keep properly hydrated" instead of "drink enough"; "maintain a proper fluid level" in place of "don't get thirsty." Despite the evident bombast, it all boils down to avoiding the plight that Coleridge's sailors encountered: dehydration.

Aside from the physical discomfort of thirst, scientists believe that dehydration may be a significant contributing factor toward getting the bends. Whether or not this is true, it doesn't pay to take chances when one possible cause of DCI can be alleviated so simply. Ergo, drink and be merry . . .

When diving for a month on wrecks in the Great Lakes I found it truly refreshing to swill down a few swallows of clear, cold water from around the mouthpiece of my regulator. It relieved that dry mouth feeling that comes from breathing compressed air from which the moisture has been filtered - good for the tanks, bad for the throat. Unfortunately, I got so used to the pleasure of nature's nectar that on my return to the ocean I forgot I was submerged in a salt water environment. I gagged when I lapsed into my four-week routine. It took only a single gulp to change my recently acquired habit.

Likewise, cave and quarry divers can sip from the bounteous supply of fresh water in which they are immersed. Those who dive in the briny deep, however, can slake their thirst only by taking their own supply.

One company manufactures a water reservoir that is placed in line with the regulator. It can supply up to a pint of water. If you're planning extensive bottom times or lengthy decompressions, you might consider investing in this device.

If you're economically minded (or just plain cheap), you can carry potable water in a plain plastic bottle. Because water is non-compressible you don't have to worry about the bottle being crushed by the depth - until you imbibe some of the contents. After the first libation either the plastic will indent or sea water will be sucked in to fill the void as soon as you take your lips off the spout.

Any old plastic bottle will do the job, such as a soda or pop bottle, but better is a bottle made of soft plastic and whose sides are squeezeable. Get one with a cap. After removing the cap take a deep breath and hold it, remove your regulator, insert the spout between your lips, and squeeze water into your mouth. (For reasons too obvious to mention, do not drink while ascending.) When you've had your fill, keep pressure on the bottle while replacing the cap so sea water won't be drawn inside. A baby bottle with a collapsible liner will also do the trick.

I've found that the first nip will taste a bit salty because of sea water residue on my lips and tongue. If you don't care for that jolt of saline solution, spit out the first mouthful around the sides of the spout, then drink as you would from a canteen or a cycling bottle with a squirt spout.

Now that we've absorbed the concept of tippling water under water, it's only a short spurt to substituting other liquids. Add sparkle to your diet with flavored soda water. Citric juices serve as great thirst quenchers for some. V-8 juice or eggnog (without the liquor) are both satisfying and filling. Or sweetened drinks such as tea, Tang, or Kool-Aid may be your preference. And if you're looking for an electrolyte treatment, Gatorade or one of the derivative athletic drinks that are chemically balanced might suit your fancy.

In short - if it's liquid you can drink it under water. It goes without saying that alcoholic beverages of any kind are forbidden. No beer or wine coolers, either.

And another word to the wise: don't guzzle too much without reading the section on the potential urinary consequences. And stay away from strong diuretics such as coffee.

Provender

"Hey! Are you serious? You can't eat food under water." Wrong! I can, and I have.

In these days of long dives and longer decompressions, maintaining stamina has become an issue of note. Domino's doesn't deliver under water, where fast food is defined as a quick bite from a speeding shark. Dining during a dive is not intended primarily to stoke the body's holds, as a seven-course meal would do, but to assuage the discomforting pangs of hunger so you can concentrate on more important matters at hand, and to supply quick energy to replace that which was lost during exertion in frigid water.

The types of food under discussion here are more properly referred to as snacks.

I once took down a leftover soft-shell crab in a sealed plastic storage bag, but it was strictly for a gag. After climbing onto the boat I reached into my mesh accessory bag and, unnoticed, unsealed the plastic bag to remove the stretched-out crustacean. I then pulled it out and displayed it just long enough for people to see what it was. Naturally, they assumed it was alive. When I took a giant bite out of its head, left a gaping wound in its body, and turned around with claws and meat hanging out of my mouth, three people immediately rushed for the rail and threw up. I was not a popular person after that.

I haven't yet found a way to get a ham and cheese sandwich out of a plastic bag during a staged decompression, and get it into my mouth before the bread dissolves to salty mush. (Although, the consistency of the ham and cheese remains the same.) The problem with most comestibles is that they are no longer palatable when gobbled with sea water. Furthermore, solid foodstuff needs to be chewed: an action that is difficult when you've got a chunk of rubber between your teeth.

Difficult, but not impossible, as I've learned. Here's the secret. Stick to insoluble victuals and fare that won't become impregnated with water. Beef jerky comes to mind because I've tried it during a hang, but pre-cooked hot dogs would suffice as well. Hold your breath, remove your regulator, tear off a bite of the meat and stow it in your cheek, replace the regulator, and resume breathing. Now slip your teeth off the mouthpiece but retain a grip with your lips. With a little practice you'll learn to breathe and chew alternately. Be very careful swallowing.

Once you've mastered the technique you can try apples and pears and other fresh fruits. You may even add vegetables to your underwater cuisine. Naturally, you'll be pre-occupied during the bottom phase of a dive, but deco time is chow time when there's little else to do.

Another trick I've tried with success is reusable plastic squeeze tubes. I generally use these for climbing, backpacking, and wilderness canoe trips. The tube looks like a toothpaste tube: one end has a screw cap but the other end is open. The open end is sealed with a key similar to the kind used to get all the toothpaste out of the tube: it is wound up to force the contents out the neck.

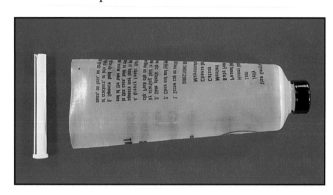

These tubes can be packed with jelly, applesauce, or any spreadable edible of similar consistency. Peanut butter does not work well because it sticks to the inner surfaces of the mouth and takes too long to ingest; it also gets all over the mouthpiece and creates an awful mess. You need grub that can be swallowed between respirations without leaving remnants that might interfere with breathing or be inhaled into the lungs.

Brand name products such as Cheese-Whiz can be heartily consumed. Don't take it down in its original pressurized can because the hydraulic force of the water at depth might cause the container to implode and expel its contents catastrophically. There's nothing worse than a mesh bag full of cheese shrapnel. At home, squirt the processed cheese into a reusable squeeze tube instead.

To get pasty foods out of the tube simply unscrew the cap, insert the end of the tube into your mouth, and suck out the contents while squeezing the tube. Make sure to replace the cap before releasing your grip, or water will be siphoned back into the tube.

You might not satisfy the appetite of a connoisseur, or get sustenance from all the major food groups, but you'll at least quell that grumbling in the pit of your stomach that is keeping the other divers on the deco line awake.

The future has a need for good underwater chefs.

To Pee or Not to Pee

Surely you jest, you are probably thinking; and surely I will. But although I treat the subject lightheartedly there's a serious side to the matter that cannot be eliminated. In Victorian circles the unpleasant realities of life are couched in euphemisms so vague that one often wonders what people are talking about. But in continents where cultures are more open to discussion one is more likely to hear, "When ya gotta go, ya gotta go."

> In days of old, when dives were cold,
> And drysuits weren't invented,
> The best latrine was neoprene,
> Especially when rented.

The standard catechism in diving is, "Have you ever peed in your wetsuit?" To which the response is either a truthful "Yes" or the embarrassed falsehood "No."

With the advent of the drysuit, however, piddling under water became less fashionable. No longer could a diver feel the warm flush of fluid that temporarily dispelled the cold. He perceived instead a spreading puddle that saturated the fabric of longjohns and insulated jumpsuits and created a clinging, soggy mess.

I have never "gone" in my drysuit, but I've come pretty close on occasion. The experience was always a painful one. I couldn't even run around in circles to hold it in. I just had to suffer while my bladder threatened to explode.

Thus arose the era of limiting intake and fighting to "hold it in." This system works okay on descents of short duration, but with the evolution of technical diving and considerably longer immersions, the former scheme is unwholesome and can lead to medical complications, and the latter is often impractical. Doctors agree that it is more healthful to drink till your cup runneth over than to alter the body's chemical balance through enforced dehydration. There has to be a better way of reducing urinary pressure than by getting out that last squirt before zipping up the suit.

And there is. In fact, there are several alternatives. So let's accept the inevitable and examine the various means of "going" when the necessity is pronounced, and in ways that are both convenient and, among divers, socially acceptable.

Diapers

The number one solution to the problem of a distended bladder is also the oldest. It involves an apparatus with which every reader is not only familiar but in whose management he has already excelled. Before you start groaning at the mention of the soft absorbent material known as diapers or nappies, understand that I'm not talking about Pampers or dydees for infants, but about the extra large version sold for adult incontinence.

Loss of bladder control and lack of intestinal fortitude are facts of life in hospitals where patients are elderly, unconscious, bedridden, or confined to intensive care units. For that reason there are a number of manufacturers that market a variety of devices to ease the inconvenience of the natural order of events. Their motto: what goes in must come out.

I hardly need tell you that insulated underwear soaked with ammoniated water loses a certain amount of thermal value even if the liquid has been prewarmed. Polypropylene may wick away moisture, but it can't wick it very far inside a drysuit. You're stuck with it because the evaporative process is restricted by enclosure. Worse yet is having all that loose urine sloshing around inside the lining. And while an ounce of prevention is worth a pound of cure, sometimes you just have to go with the cure.

Adult devices work the same as the baby counterparts. Naturally they are bigger to accommodate larger bodies, and have more surface area and thicker padding. Fortunately they are disposable. No running to the Laundromat after every dive.

My friends all laughed with I first donned diapers before a long, deep dive and lengthy decompression. Now the humor is gone because diapers are all the rage. Today no technical diving activity is conducted without some of the cast wearing diapers and, if they're smart, plastic panties for certain containment. This uncustomary loincloth has even gained a modicum of dignity.

Diapers are readily available in supermarkets, pharmacies, and convenience stores, and are inexpensive to buy. Unless you do an exorbitant amount of diving and subscribe to a diaper service, paper disposables are more convenient than cloth reusables. Simply toss the soiled diaper in the trash at the end of the dive, even if you didn't "go" in it. The plastic panties can be rinsed and reused.

As with any new piece of gear there are a few tricks to learn in order to become proficient with its use. (I will assume that you have long been out of practice with proper diaper etiquette.) The absorbent material is surrounded by nonpermeable plastic designed to prevent partial spillage as a result of overindulgence.

Before pumping bilge in my drysuit and discovering too late that I had exceeded the diaper's capabilities, I ran a few tests in my bathtub: wet runs, I guess you could call them. It's a good thing I did.

Every morning for a week I stepped into the bathtub immediately after awakening. I donned a diaper and peed. When the experiment was over, I removed the thoroughly soaked appliance, soaped up, and rinsed off under the shower. I learned a lot.

You can't let it all go in one continuous spurt. The diaper floods so rapidly that the urine cannot soak throughout the absorbent material fast enough: it's like pouring a glassful of water over a sponge instead of slowly inserting the sponge into the glass of water. The result: leakage around the plastic barrier and wee-wee dripping down my leg.

For optimum efficiency use short, controlled bursts. And don't wait until you absolutely *have* to go. If you do, your bladder will erupt like a burst dam and instead of irrigating the surrounding fabric you will inundate the material with considerable dismay. Let urine trickle out as you feel the need.

Diapers work better for women than they do for men because female directional parameters are naturally optimized. Due to the construction of the male anatomy, it's necessary for men to position their equipment so there is no restriction in flow (watch out for crotch straps) and so the angle of discharge is favorably aligned. A vertical orientation causes a hydrant effect that results in abdominal saturation. I've found that overall absorption is more even and efficient when the diaper is hiked up in front where it's centered over the sluice gate.

After you wet your diaper you might acquire a skin condition that babies commonly get: diaper rash. A few shakes of talcum powder will minimize the problem. Sprinkle powder in the diaper before you pull it on; this will absorb perspiration and make you feel more comfortable even if you don't pee. If you do take a whiz, get out of the diaper as soon as you can, rinse your bottom with a damp cloth, dry yourself thoroughly, and pat powder on the affected epidermis.

Two unspoken rules: never ask your buddy to change your diaper, and never offer to change anyone else's.

Catheters

More sophisticated than diapers are external containment systems. For women there is a female incontinence management appliance consisting of an elongated rubber form-fitting cup which seals against the skin and which is secured firmly in place by means of straps around the waist and thighs; an extension tube assembly; silicone lubricant; and outer support pants. The soft rim of the cup, where it presses against the flesh, must be coated with lubricant to ensure a good seal.

This is actually designed for non-ambulatory patients, so how effective it is for athletic activities I can't say. I haven't been able to convince any women to field test it in my bathtub. Perhaps I could get Urethra Franklin?

For men there is the condom catheter. Advertised variously as twist proof, extended wear, and self-adhesive, the device is similar to the contraceptive and safe-sex counterpart with some exceptions. Besides the fact that the latex is not ribbed, the condom can be secured with clamps, straps, adhesive strips, or glue (not Superglue). You'd think that similar precautions would be taken to prevent pregnancy.

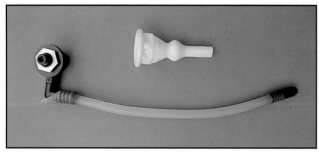

The male urinary catheter further differs from the birth control device in that the condom has a hole in the business end. You don't want to get them mixed up and grab the wrong one from the medicine cabinet in the middle of the night during the throws of anticipated ecstasy. You

Wrecks such as the *Wilkes-Barre* provide underwater rest room facilities, but you have to be quick on the draw in order to prevent flooding and backwash.

might get unlucky. They're sold in starter kits that come with twelve condoms plus accessories (not marital aids).

Men with frail egos beware. Condom catheters are sized for small, medium, and large. If you buy one that's too big for your hardware, the condom will not fit snug on the lesser diameter so you'll leak around the seams. The sizes are color-coded, too, so there's no way to disguise the caliber of your gun. What you see is what you got. And what everyone else sees.

The female appliance is washable and recyclable but the condom is not. The inner wall of the condom is generally coated with an adhesive that sticks to the skin as the material is unrolled; without self-adhesive, a special glue must be applied during emplacement - a sticky job at best. For additional safety precautions against premature divestment during the venting procedure - pissing off the condom, so to speak - some brands offer clamps or straps which are fastened to the anatomy. Most divers consider this additional security an encumbrance.

In addition to hospital applications, both sexes of external catheters receive widespread use in military flying and the space program, where pilots and astronauts are forced to spend many continuous hours in a cockpit or control cabin. I don't know what modifications have been made for waste management at high altitude, because as far as I know, this technology is not available to the public.

If you're truly macho or hard-core you might wish to insert an internal catheter, which can be purchased from a hospital supply company.

Storage Facility and Venting Capability

Neither the male nor female catheter would be of any value unless it siphoned or funneled the discharged liquid to a suitable location where it wouldn't drench your clothing. Two methods are available to achieve this goal. Both employ a length of flexible tubing which is secured to the outflow end of the condom or to the drain spout of the crotch cup, and which connects to a storage or elimination device.

For general incontinence there is the reusable liquid-tight bladder or sack which is strapped to the upper leg and which is similar in appearance to a hot-water bottle. The contoured drainage bag is fitted with a non-return valve which prevents urine from flowing back into the tubing when you become horizontal or inverted, or when ambient pressure becomes sufficiently high to squeeze liquid out of the bag.

Now you can dribble whenever it's convenient and without thinking much about it. Urine flows through the tubing into the bladder until it swells to capacity - generally about a quart, but larger bags can hold as much as thirty-two

The external assembly of the p-valve.

ounces. Don't exceed the bladder's full measure or the back pressure might burst the condom or force urine past the terminal roll, or break the seal of the cup. A soggy aftermath will ensue.

Empty the bladder after the dive. Both it and the tubing can be rinsed and reused.

An appliance that was developed specifically for diving is the drysuit port or p-valve. The valve is installed in the crotch area of the suit by cutting a hole through the material. Inner and outer plates are both o-ring sealed. An external screw cap protects the outer valve assembly from salt encrustation and mineral build-up which might eventually cause the valve to leak as much as the user - an undesirable side effect.

To relieve yourself, unscrew the safety cap and let loose. Bladder pressure forces urine first through the biological valve and then through the mechanical valve into the surrounding water. A one-way flap mechanism prevents ambient pressure from reversing the flow. For insurance purposes, replace the external screw cap after each spurt. Flushing the tube and valve assembly with fresh water prepares them for another dive.

Whatever method you choose, being prepared for the unavoidable event often instills a psychological advantage that obviates the need: when you know that you can "go" whenever you have to, you often don't have to.

Spartan accommodations on the wreck of the *Vienna*. The crescent moon is an age-old symbol signifying toilet facilities.

The End

Thus far there has been no movement to contain the harder reality of excretion. In that regard, all I can suggest is that the diaper is a fundamental appliance that "covers all bases."

But who knows what they will think of next?

UNDERWATER COMMUNICATIONS

If you think people have communication problems today, you should have been around a couple of million years ago.

Man's earliest form of information exchange was a miscellany of grunts, waves, and physical contacts. By this method the cave man was able to limn for his fellow cave men and women some of what he observed about the world. He learned to describe objects and to portray actions of immediate consequence in a rudimentary fashion that enabled him to survive in a primitive and often hostile environment. Long before *The Elements of Style*, people strove to articulate thoughts in a manner that they hoped would not be misinterpreted.

When life was simple, a shout of alarm and a finger pointing frantically in the direction of a skulking saber-toothed tiger was all that was required to convey a sense of danger. A touch could call attention. Facial gestures could express most feelings. As cultures evolved and tribes grew into nations, the proper exchange of ideas became necessary for racial survival. Thus came speech and the invention of language.

Nowadays, due to the complexities of modern society, verbal intercourse is much more complicated than it was for the ancient troglodyte. Linguistics has become a dedicated science, gossip an art, and rhetoric an end unto itself. Yet, despite the meticulous pronunciation of words and the strict usage of grammar, conversation is often misleading.

Personal relationships can be severely strained by mere inflection or tone. Wars have been started and millions of people have died due to seemingly minor misunderstandings. A flat statement's meaning is sometimes open to conjecture, depending upon the listener's state of mind and what he expects to hear. Written concepts can be lost in translation.

Notwithstanding the above, divers daily thrust themselves into an environment more hostile than any ever encountered throughout the evolution of the species. In a medium where death is literally a breath away, precise communication is vitally important. So by what inspired mechanisms do divers convey their thoughts and transmit messages under water? Through a miscellany of grunts, waves, and physical contacts. Mime has come full cycle.

Sign of the Times

To overcome some of the distortions of meaning apt to occur in such an indefinite means of delivery, a system of digital gestures has been standardized whereby a diver can report his condition to people in the vicinity. Banging on your tank with your knife will attract attention, but very little else; it is equivalent to the cave man grunt. Swinging your light rapidly will accomplish the same goal, as will sneaking up on your buddy and grabbing his leg from behind. (Although, the response is not always the one anticipated.) There has to be something better.

In the early days of diving when hardhat prevailed, the diver employed a technique that was similar to Morse code in order to signal his topside tender. Assuming the tender was awake, the number of yanks on the hose could mean either hello, good-bye, or get me out of here. (British tenders never had yanks on their hoses.)

With the introduction of American Sign, divers began to realize that life had more meaning with hand signals. Since most people have grown up making rabbit impressions on lighted walls by manipulating their fingers in front of a projection lamp, hand signals at first glance seem to be the answer to underwater communication. Entire books have been written about it. I advise every diver to learn and use such signals. However, you should be aware that the systemization of hand signals is subject to dialect the same as the spoken word.

For example, during a cave dive I thought I heard bubbles escaping from my tanks. I waved my light at my buddy and jerked my thumb over my back. Since in the cave diving lexicon the curled fingers and extended thumb signify the termination of the dive, he interpreted my meaning in light of the thirds rule, assumed that I had reached the turnaround point on air supply, and began to exit. Because there's no explicit way to semaphore the equivalent of a pencil eraser, it took quite a bit of waving and gesturing to get the correct point across and convince him that it was okay to continue with the dive. A week later I used the same fist with outthrust thumb on the deco line to express enthusiasm to my wreck-diving buddy over the great dive we had.

This kind of flaw in the language of manu-communications is equivalent to spoken and written words like "run" which, according to the dictionary, has more than one hundred definitions, each dependent upon the particular usage, context, and grammatical construction.

In addition to these colloquial differences between diving communities, there are local variations among shops, clubs, or closely associated groups of friends who have developed their own vernaculars. If you're used to diving off the New England coast, and go on a trip to southern waters, you're likely to encounter a drawl that you might not recognize - or worse, misconstrue in times of emergency.

Don't Take Slate for Granite

Paralleling the evolution of cave diving with cave living came the invention of the slate. The modern day message slate is made of plastic instead of metamorphic rock, but "polymerized organic compound" is too long and cumbersome a description even if it is more accurate. Substitute black insoluble graphite for chalk, and what could be better than the written word?

At first heralded as the crowning solution to

Above is a slate that is clipped to a snap hook. To the right is a slate that is mounted on the back of a gauge panel.

posting flash announcements, and only slightly less effective than syndicated service, certain unanticipated difficulties accrued that made slate writing (or slating) not always effective.

If you're the kind of person whose handwriting deteriorates rapidly when your fingers get cold, or when writing on a hard surface without a blotter or several sheets of paper to smooth out the jerky motions of that cheap ball-point pen, or when scribing with an unsupported wrist, think about how arduous it is to concentrate on calligraphy at a depth of 200 fsw, or how awkward it is to print simple block letters while being thrashed by the anchor line during decompression in six-foot seas.

Don't get me wrong. Slates have a definite place in underwater communications, depending upon the circumstances of the dive. Use them when necessary to exchange information with other divers, to send dispatches to the surface, or to record numbers or observations you don't want to forget.

The curved wrist-mounted slate.

But if you see a shark about to make lunch out of your buddy: grunt, bang your tank, flash your light, wiggle your fingers, shake your fist, and hit him with your slate - just don't take the time to write a biological treatise on stereotypic premastication labial movements of the Carcharodon carchias. The lips you lick may not be your own.

Tell it Like it Is

Then along came VOX - a disturbed acronym for "voice activated transmission." (VAT is more acronymic, but vox is Latin for voice.) This does-n't mean that you can tell your car to change gears by talking into the steering column, although that technology will undoubtedly arrive some day. VOX is a wireless communication device that allows two divers to converse during the course of their dive, both with each other and with top-side personnel. It's a bit more sophisticated than holding your breath while your buddy screams through his regulator into your ear, and a whole lot more understandable.

In its simplest form, VOX consists of a surface unit that drives two transmitters and two receivers: one set for you and one for your buddy. (If you had only a single set, you could do nothing more than soliloquize.) Hardwired versions are available, but once committed to a tether, a scuba diver no longer has complete freedom of move-ment - like two kids with a string and a couple of Dixie cups. Each diver carries a transducer that exchanges signals with a transducer/receiver lowered into the water from the surface unit.

In some models the earphones are a pair of speaker enclosures worn over the ears like the stereo headphones of a Walkman. The electronic circuitry and batteries are housed in the enclo-sures and hardwired to a head-mounted trans-ducer (or 'ducer, as it's called in the biz). Wires are external, protruding out of the speaker enclosures like spark plug wires from a distributor cap.

For that reason there might be some concern about snagging the wires on protruding rocks or jagged wreckage, or with dislocation due to con-vulsive contact with a bouncing anchor line. On the other hand, because the earphones are enclosed, extraneous ocean noises are practically eliminated. The antenna can be positioned down along the side of the face where it's out of the way.

Since sound in water is received directly through the skull instead of through the tympa-num (the vibrating membrane known as the eardrum), some models incorporate what are called bone oscillators, or bone conductors: coin-sized pads placed against bone in the vicinity of the ears. They can be attached to the mask straps, held in place by a head band, or taped directly against the skin.

Some units are sensitive enough to be worn outside the hood, although in practice the recep-tion is not as good. Voices tend to come through muddled because of air trapped by the hood and, if the hood is made of neoprene, because of bub-bles trapped in the material. The low profile is particularly advantageous for wreck diving appli-cations.

For distances up to half a mile the transmit-ter/receiver can be a small encapsulated sending unit powered by rechargeable ni-cads as small as AA. Since most wreck sites are confined to an area smaller than a city block, this might seem suffi-cient. But another function of battery power in addition to signal strength is the duration of oper-ating time.

(Before you exclaim that even the least intre-pid cave diver would soon be out of range with such a device, let me offer the caveat that the sys-tem won't work through rock. It is primarily a line-of-sight instrument which can penetrate a few bulkheads but works best in open water.)

For increased range and extended use, a box-encased battery pack and sending unit can be mounted on a tank (a scuba cylinder, not a fighting machine), backpack, arm, or any convenient belt or strap. Batteries are either rechargeable or can be replaced in the field by the user (out of water, of course). If you can't rely on having onboard alternating current, or are exploring caves in some forgotten corner of the Yucatan Peninsula, carry spare batteries.

The transducer is an essential element that is no larger than the average wrist-worn compass. In compact VOX units the transducer is connected to the electronic components by means of a waterproof extension cord so it can be positioned conveniently. One type is designed to be worn atop the head like a truncated rhinoceros horn. It may look silly but it is highly effective, providing all-around transmission and reception.

Jamie Powell with an arm-mounted unit.

In larger systems the transducer is attached directly to the pack containing the battery and electronics, eliminating the extension cord and keeping the transducer in place and out of the way. However, transducers worn on the tank can create a problem known as "shadowing." If you face your partner during a conversation, the signal line-of-sight goes through your tank, body, and BC. Metal and organic density can affect acoustic quality to a minor degree, but the real culprit is air in the BC or drysuit. Phase shifting caused by the signal cutting through the air/water interface disrupts the harmonic rhythm and breaks up the smooth, continuous flow of sound.

What you get is interrupted communication as your partner's voice fades in and out. You have to turn sideways and look at him askance, as if you were cocking your ear, in order to understand what he is saying or to be properly heard. One solution to this kind of interference is to mount the unit high enough so it peers over your shoulder. The problem can also be rectified by adding a boom cable so the transducer/receiver comes up behind your head. (This is not meant to imply that technical divers have nothing between their ears, or that they don't have enough brains to stop an electronic pulse.)

I Have no Mouthpiece, and I must Scream

The last item of equipment necessary to implement speech is an air chamber in which to speak. You can't talk with a rubber mouthpiece jammed between your teeth; at least, not very effectively. Enunciation suffers when the tongue and lips cannot cooperate to articulate certain consonants. If you can produce only vowels intelligibly then you're back to the cave man grunt.

The easiest way to get your lips and tongue back together is to pull out your regulator and surround your mouth with a rubber pocket into which the transmitter is plugged. This simple device looks like a flattened Dixie cup and serves the same purpose. It's called a half mask or mouth mask, and fits comfortably below the face mask that covers your eyes and nose.

So you won't be limited to sentences that can be uttered only one exclamation at a time, and to preclude having to switch back and forth between the regulator and the cup (or to have to choose between breathing and talking), the mouthpiece is removed from the regulator (prior to the dive) and replaced with the mouth mask. The mouth mask seals around your lips and is held in place by a head band.

The regulator works on demand as usual, the only difference being that instead of having a mouthpiece holding your jaw open, the half mask presses against your chin and seals around the mouth only. The system is universal so that any regulator can be used.

People who use a mouth mask complain of jaw fatigue and a leaky lip seal. Whenever you talk you must move your jaw, an act that stretches the band holding the cup to your lips. And unless you're pudgy-cheeked, the bony turn of your jaw will create tiny channels that quickly fill with water. Not to fear: the chamber can be emptied by a strong expulsion of air the same as if you were purging your regulator.

Another disturbing feature of the mouth mask is the snorkel effect. When you look up and inhale at the same time, the water column distance between your lungs and your regulator is lengthened by several inches. You have to inhale hard the same as if you were sucking air through the tube of a snorkel. It doesn't sound like a big deal, but the difference in ease of breathing is noticeable.

Full-face Mask, Again

The obvious answer to the cumbersome difficulties of the mouth mask is to wear a full-face mask with an integral regulator and built-in microphone. In fact, most commercial communication units are designed for such use. Since I have already described the intricacies of the full-face mask in the previous chapter, I won't repeat the information here. Suffice it to say that the performance level of a comm-unit is enhanced by using a dedicated system. You can breathe and talk in comfort the same as if you were sitting in your parlor. And in case you're wondering, the full-face mask can still be taken off under water without harming any of the electronic components. They are individually sealed.

Communication Difficulties

Although the word VOX is on the way to becoming generic, some communication units do not incorporate voice activation, nor does voice activation work all the time. Push-to-talk back-up is an important option. Having to press a button every time you want to talk obviates one of the beauties of the VOX system: hands-free operation. But even that is better than having no voice at all.

What is it that inhibits voice activation? Largely, air density. As you dive deeper the pressure increases and your tone of voice changes even if you aren't angry. At 130 fsw you might find yourself groping for manual override and revert to hand signals.

After about 220 fsw pitch and tone are so altered by the thickness of the air that speech becomes distorted and words become difficult to understand. At any depth it is important to practice proper enunciation. Use simple words and short, curt sentences. Exclamations have more impact than explanations. Learn to avoid multi-syllabic constructions, long descriptions, and words whose pronunciation is ambiguously similar to that of other words.

Don't get short tempered when your partner keeps saying "Repeat!" Remember that he may have received your message during an exhale, when the noise of exhaust bubbles was added to the background garble of muffled transmission, clanging gear, and ocean noise.

Worse yet is trying to communicate through VOX when breathing helium as your inert gas. Not only does sound propagate with increased velocity in a helium atmosphere, but there is a change in the normal frequency of the vocal tract. We are all familiar with the Munchkin voices of deep saturation divers who need to speak through an descrambler in order to make themselves understood. The best you can do with VOX is to lower your voice and speak slowly.

Most people love to talk but are very poor listeners. Instead of hearing what is being said, they lie in ambush for an opportunity to butt in with comments of their own. As with any electronic communication device - whether it be a CB radio, car phone, or single-sideband radio - it is important to practice good listening habits.

Communicating through VOX in shallow water does not present the problems that are associated with depth. Sounds are sharp and clear, modulation is distinct, interference is minimal. About the only frustration you'll experience is that it becomes impossible to whistle deeper than about 50 fsw. So much for passing that boring decompression to the tune of "Bridge Over the River Kwai."

Serendipitous Benefits

On the other hand, the surface unit comes with several options that can add to your listening pleasure. For one thing, you can record your conversations and play them back later in order to reconstruct your dive. By timing the replay with a stopwatch and keying off specific descriptions, you'll know where you were and what you were doing during the course of your dive, how long it took to swim from one spot to another, and the length of time required to perform a specific task.

You can also use VOX not only for communication with a buddy, but to record observations. No more time-consuming scribbling on a slate or relying on memory to take notes afterward in

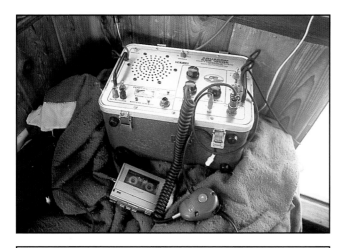

Topside unit and cassette recorder. VOX equipped divers can communicate only between themselves unless a topside unit is added to the system. By submerging a transducer connected to the topside unit, divers can exchange conversation with topside personnel (notice the microphone). An inline cassette recorder can then tape all communication for posterity.

order to reconstruct the sequence of events. This is quite an advancement compared to the ancient system of record keeping and dive log entries.

Need I bother to mention how easy it is to call for help or to describe a perilous predicament? Instead of gesticulating wildly you can calmly ask your buddy to see where that air leak is coming from, or tell him exactly where to look for that entanglement pulling against your left side. You can even ask topside to send down certain tools with the next team, in order to work more effectively on a task at hand.

Some comm-unit models have an optional video camera receiver that plugs into the underwater housing and records conversations while you shoot - no more listening to those stupid bubbles on your home videos. For that once-in-a-lifetime shot of your Great White shark encounter, even if you and your footage become a snack, the dinner crowds can still be enthralled by your last gasps, over and over again. And all at an affordable price.

Music to my Ears

Once I was diving deep inside a wreck at 130 fsw, moving forward cautiously through the darkness and stirred-up silt. I came to a hatch leading into another compartment, and just as I was contemplating on whether to penetrate farther, I heard Sonny and Cher singing "I Got You Babe." It was the clearest stereophonic sound I have ever heard - as if the duo were standing right there beside me. I got out of there so fast that suction

pulled out all the fish as well.

I didn't know what was going on. Were the voices from heaven, angels come to guide me to that great shipwreck in the sky? No, it was someone on the boat playing a tape recorder and piping the music into the water through a submerged transducer. Once I figured it out I enjoyed the melodic decompression stops, but I wish I'd had a warning.

Steve Berman has gone a step further. He sealed an audio cassette player in a watertight container, then ran wires through the plastic bulkhead to earphones dipped in rubber. During lengthy decompressions he listens to his favorite tunes. When he demonstrated for me his underwater counterpart of the Walkman, I promptly dubbed it the Diveman!

For longer playing time he ran leads from the battery terminals to larger, external batteries. The cassette cannot be changed or reversed underwater, so the tape plays the same side over and over.

Parting Disclosure

When all is said and done, underwater communication devices have numerous advantages and no disadvantages - unless you dive with a chatterbox or gossipmonger. Once you start using VOX you'll never want to be without it. To understand what it's like to be dumb, try going into a store and describing what you want with grunts, waves, and physical contacts. Then, with your imagination, compare that experience with what you could have accomplished in the water should you have been able to communicate instantly and understandably with your buddy or with other team members who were wearing comm-units.

The simple act of describing what you are seeing forces you to concentrate your focus. It enhances your memory of the moment, and heightens your awareness of the experience while saving your on-the-spot impressions for posterity. Communication is more than a way of saying hello to your buddy, it's a means of staying in touch with yourself.

My advice? Put an end to the silent world.

DIVER PROPULSION VEHICLES

It was shaped like a torpedo with a rounded nose at the front and a propeller at the back. Two men rode it astride like cowboys on a galloping horse. One of them controlled the throttle with a joystick as the submerged vehicle sped through water blackened by night. The waves whipped by so fast that the frogmen had to hang onto their masks lest they be torn off their faces by the jetting sea.

Their objective was in sight under a moonless sky. Flickering starlight guided them toward the gray silhouette. As the hydroplanes turned downward, the vehicle slipped beneath the surface like a silky sibilant wraith. Now only a compass pointed the way. They maintained their bearing with heraldic intent until the shadow of the great ship loomed overhead. They did not announce their arrival; not then, anyway. Not until they abandoned their vehicle and slipped quietly ashore did anyone discover their presence. By that time it was too late. They grinned solemnly over the accomplishment of their mission.

. . . Each was shaped like a torpedo with a rounded nose at the front and a propeller at the back. One man rode each astride, like two bikers hunched over the handlebars of their machines in a neck-to-neck race along a bright cerulean track. They plowed through crystal clear water which glimmered with wavering sunbeams that knifed downward like pirouetting klieg lights in a modernistic stage show. The reef was in motion with many-hued tropical fish, soft corals and sponges, and sea fans that waved like beckoning Sirens.

The frogman took the lead. The reef passed beneath the vehicles in a rainbow montage. A sunken boat loomed up out of the sand, its awkward lines offensive against the natural curves and gentle shapes portrayed by nature. The motors whined down. The frogman got off his T-bar seat and investigated the mortal remains of a fishing scow; the other nodded in bewilderment. Then, machines between their legs, they pulled back on the hydroplane levers and rose to the surface of the sea, where they lolled in the noonday sun and waved slowly to attract the attention of their chase boat. Both men rejoiced over the accomplishment of their mission.

These two scenarios took place in different oceans that were thousands of miles apart, were separated in time by a span of fifty years, and were distinguished by man's penchant for war and peace. The only common thread in the stories is the underwater vehicle: the former was called a chariot, the latter a diver propulsion vehicle, or DPV.

The two Italian frogmen who rode the infamous chariot were among a handful of intrepid undersea warriors who slunk beneath the cool Mediterranean in order to sabotage British warships that the Allies had overconfidently believed were docked in safe harbor. World War Two was in full swing. The load of explosives laid under the enemy keel broke the ship's back and sank it, right under the noses of unsuspecting British tars.

The chariot was actually a steerable torpedo: a high explosive warhead to which was affixed a battery powered motor in a sealed compartment, and which was directed by a two-man crew wearing oxygen rebreathers and full-face masks. The chariot was *not* a miniature submarine; the men straddled the torpedo behind a window screen but were otherwise exposed to the elements. Their feats were fraught with daring.

The frogman in the second scene was my

longtime friend and fellow underwater explorer Billy Deans, also known as the Frogman. I was his dive buddy on an endurance run off Key West, Florida. We were testing a DPV imported from Germany called the Aquazepp. I jokingly called his machine the Aqua Zeppo, and mine, his silent partner on this trial, the Aqua Harpo.

Two generations separated the chariot from the Aquazepp. The evolution from military machine to civilian excursion and transport vehicle occurred slowly. As always, demand necessitates invention; then, adaptation begets clones. Along the way were some offshoots that more rightly deserve the name "scooter," although in practice many of them never see service in that mode.

The Underwater Scooter

During the proliferation of motorized propulsion vehicles came the popular hand-held type whose only controls were an on-off switch and a two-speed throttle. A watertight compartment the size and shape of a fat watermelon housed a motorcycle battery and an electric motor. The propeller was protected by a shroud which not only prevented a diver's "danglies" from fouling the prop (and severing his hoses) but which funneled the water into the blades thereby increasing the propeller's thrust: instead of some water being pushed wastefully outward, it is all forced backward. That is the reason the tolerance between the shroud and the flattened edge of each blade is so critical. A slight deformation of the shroud will drastically affect the performance of the vehicle.

Some units have a headlight in the nose cone to blaze the trail in front, or to obviate the need to hold and aim a light by hand when all your fingers are fully occupied by hanging onto the scooter's handles. Thus, when charging along the corridor of a submerged cave or descending upon a wreck, the diver can aim his scooter toward the object of his delight and see what he is zooming in on. If you release the throttle control and hover to a stop, you can point the scooter like an oversized dive light.

There are a few things to know about effective scooter operation: things that are not included in the instruction manual but are picked up in the field. (I once had a college professor who said, "Whatever I don't cover in class will be covered in the test.") Why reinvent the wheel?

Care and Safe Stowage

Because a scooter is an expensive piece of equipment it makes sense to protect it from damage or loss. By this I don't mean increasing the comprehensive on your homeowner's policy. The best insurance is preventative. Most people I see with a scooter lay it on the deck of a boat and forget about it. A few with more forethought will tie it to the rail with a lanyard. Usually, this only confines the scooter to an arc described by the length of the line.

At the first roll of the boat the scooter performs its primary function by scooting across the spray-slicked deck into a set of steel doubles, a bout that the scooter is bound to lose. Then, as the boat settles back on an even keel or rolls the other way, the scooter follows suit. It either wallows back and forth like the watermelon it resembles, or, if placed upright on its shroud, promptly tips over.

Lash your scooter securely. Once a boat begins careening in heavy seas and taking spray over the gunwales, your friends will not be inclined to help secure your scooter for you. As they huddle in the cabin, you'll find yourself battling the wind and the waves in order to save your investment from damage.

If a scooter topples or caroms off tanks and bulwarks, a plastic handle may snap off or the shroud may get permanently bent out of round. And what people seldom think about is what all that sheer stress is doing to the seals and o-rings. A scooter is an instrument, not a bean bag.

By the same token, don't toss a scooter onto the bed of a pickup and then bounce along a backwoods road at breakneck velocity on the way to a cave as if you were running a steeplechase. At the very least, wedge the scooter tight against the sidewall with soft gear bags. Preferably, stuff some towels or blankets under the casing to absorb the shock imparted by those washouts and potholes.

Protective cases are not made for scooters, but a suitably sized cooler will work just as well. Pad the interior sides and bottom of the cooler with foam, support the body of the scooter so the shroud doesn't rest on the bottom, and you're ready to go.

The storage case will also prevent salt spray from accumulating in the narrow separation between the moveable grip and the handlebar, where it can work its way into the rotary mechanism. Usually, a sticky switch is due to sea water corrosion on exposed sliders and pinions. The entire unit should be rinsed - even soaked - in fresh water after immersion in the marine environment. With the foam removed, the cooler can double as a bathtub or rinse bucket.

Even if you don't use the scooter after a day at sea, if it lay exposed to the elements it should be soaked in fresh water to dissolve those nasty salt crystals that were deposited from the spray.

Scooters flown in caves can suffer the same malfunction from quite a different source: silt and suspended clay. If sediment is not already floating in suspension, the floor might be churned up by the scooter's whirling passage: you'll fly through clouds of disturbed alluvium on your return. This ultrafine material can sift into the rotary mechanism and jam the handle the same as salt crystals can do.

Once, on a boat, I watched a scooter that was secured by the nose grip as it rolled and ricocheted off adjacent hardware. Every once in a while it would fall just the right way to actuate the

The "watermelon" type scooter in a cave. Note the tow line clipped to the diver's belt. The white objects secured to the scooter's nose are capped lengths of PVC tubing, whose flotation provides balance for better horizontal operation - this prevents the scooter from taking a nose dive.

pressure switch. The motor raced madly without a load. If this occurs for long periods of time it could eventually burn out the motor. At the very least it will drain the battery.

If an electrical contact breaks at the solder joint or a wire is pinched or snipped - due to rough handling of the machine - the motor will stop running. Occasionally, however, a circuit will short out when insulation is scraped off a conductor and it grounds against the metal casing or touches another wire. If the battery leads short out there's likely to be a meltdown: circuits will get fried, battery acid will boil, and the gradual expansion of gases (in effect, a slow-motion explosion) will force out the seals and let in water. Take extreme care when opening a flooded scooter. Parts can go flying and noxious fumes can make you sick.

The Runaway

There's nothing more pernicious with respect to scooters than a runaway. If you think I'm joking as I've done in the past, this time I'm not. Make no mistake: freewheeling scooters that take off on their own present some very real problems, not the worst of which is the loss of an expensive piece of equipment. Prospective buyers take note, users take precaution.

A scooter can develop a mind of its own in a variety of ways. If the switch wire short-circuits during in-water use the motor will not be damaged; it will continue to run till the battery goes dead. This may not be much of a consolation as you cling desperately to a scooter that is bound to turn in circles until one of you runs out of energy. It's worse if the machine is yanked from your grip and takes off like a speeding torpedo. I know of more than one scooter that was last seen on its way to Spain or that knocked itself silly playing billiards inside a granite chamber.

A similar situation can occur if the throttle or switching mechanism jams "on" while the scooter is being flown. This can happen because of encrusted salt or accumulated silt in the rotary switching mechanism (as noted above), untimely dysfunction (is there any other kind?), poor maintenance, or any of several mechanical maladjustments. For example, a battery not adequately secured in its compartment can shift position when the scooter is banked, or can slide forward or backward during descent or ascent. The weight or momentum of a moving battery may pinch wires or pull them apart and expose the bare copper ends.

Add to these scenarios the fact that some divers have modified their units so the switch can be taped down or locked in the "on" position, thus relieving them of the strain of pressing the button all the time, and the formula for disaster is complete. This energy saving ploy is no safer on scooters than it is on chain-saws. The kill-switch mechanism was installed by intelligent design.

There is no place in scootering for underwater cruise control.

A scooter that takes an abrupt downward plunge can cause an eardrum to burst before the pilot has a chance to swallow. A scooter that suddenly soars for the surface while the pilot is inhaling might easily cause an embolism, or carry the pilot above his decompression ceiling.

Hangin' On

Scooter users who wax long about the speed and range offered by motorized propulsion, and who brag about the resultant conservation of air, usually complain about forearm fatigue. Holding onto a scooter can be equated to doing chin-ups. In promotional films about the product, a scooter may appear to fly effortlessly through the water like a marine version of Peter Pan. In reality, flying is effortless only for the machine. The diver is often hanging on for dear life as the strength in his arms slowly wanes.

One way to avoid straining the forearm muscles is to lock your elbows and let your arms stretch out straight. There's no rule that says you have to bend your arms when piloting a scooter, but many people commonly fly it that way. Only when steering must you have a bent arm position, so you can lever the machine to the side like a banking fighter plane, or nose it up or down. When you're flying a straight course, don't fight it. Relax and let the machine do the work it's supposed to do.

Some divers attach a lanyard to their scooters. The lanyard can be looped over the wrist, or clipped onto the tank harness with a snap hook or carabiner. This permits them to let go of the scooter on the surface or during decompression, and not have to worry about it sinking to the bottom.

The Tow Line

The next step in this progression is to cut the lanyard to a precise length so when it's connected to your tank harness (preferably at the waist line) it keeps you in the bent-arm position when you're holding onto the handles. Now the scooter pulls you by the harness instead of by the arms. With the big DPV's this is the only way to fly.

Technically, this precise length of rope is no longer a lanyard but a tow line. It doesn't take much foresight to perceive the inherent risk of a leash that can't be disconnected in a twinkling. If the switch jams "on" and the machine takes off like a missile, you'd better be able to disengage at once from the tow line.

A device clipped to a harness D-ring might be difficult to reach through the plethora of hoses; or once reached, difficult to actuate because of the strain on the rope, the impediment of mitts, or the inability to see the device due to its location or the lack of ambient light. A snap hook or carabiner well within viewing range is preferred.

Place a clip device on the scooter end of the tow line and clip it to the scooter. Or employ two lines: one lanyard on the scooter and one on your person. Then clip the harness line to the scooter line so that both clips are in view and either clip can be actuated. Do not use locking carabiners.

A Fastex buckle might be a better connector.

Sinking to the Occasion

You might think that having an expensive piece of machinery that would float to the surface on its own is the obvious way to prevent accidental loss. Not so with detachable dive gear. Scooters are weighted to be slightly negative.

If you wanted to leave the scooter on the bottom while you went exploring without it (say, to look inside a wreck), you wouldn't want to come back out to find it gone to the surface and adrift in the current. Likewise, if you left your scooter on the floor of a cave while you checked out a small side passage, it might get damaged when it banged into the ceiling.

You can adjust a scooter's buoyancy, however, for special applications. Strap on a lead weight or a Styrofoam block, as needed. For permanent lift, short sections of PVC pipe capped at both ends make admirable, noncrushable flotation devices, and can be used to offset other, negatively buoyant equipment attached to the scooter.

Emergency Flotation

How can you safeguard a scooter should you be forced to abandon it under water? Whatever you do, don't leave it lying loose on the bottom unless you're exploring a submerged cave. In the open ocean, it might not have enough negative buoyancy to stay in place against strong, itinerant currents. If you tie it to protruding debris the surge might bash it to pieces. Try wedging it in the wreckage or slipping it under a hull plate.

The primary reasons for leaving a scooter behind are because either it stopped working or an emergency arose which required your immediate departure for the surface. A scooter left in a cave can always be picked up later; generally it won't go anywhere on its own. A boat dive presents an entirely different problem. In the latter case the scooter is not only abandoned in an unpredictable environment, but you may not be able to get back to the wreck for weeks, perhaps months.

If at all possible, it behooves you to send your scooter to the surface even if you can't go with it. This is where a Tank Inflatable Retrievable Device can save your asset. A rolled liftbag fastened to the scooter with a stout line and lashed in place with a strip of inner tube makes a dandy TIRD. You just yank the liftbag free from its lashing, inflate it from your tank via mouthpiece or liftbag inflator, and let the scooter rise to the occasion.

Note the liftbag secured to the shroud. The tow line consists of a pair of white nylon ropes, of which one is visible. The video camera mounted on top can record the entire dive. The camera is trained by pointing the DPV.

A large volume pop-buoy with a self-inflating CO_2 cartridge will work if the water isn't too deep, and if you can find one in a garage sale or a scuba club swap meet. It can be actuated faster than a TIRD.

Perhaps a combination of both devices is better yet. First you pop the buoy, then, if there's time, inflate the liftbag. This way, if ambient pressure won't let the CO_2 expand enough for the pop-buoy to float the scooter off the bottom, the liftbag can provide the initial lift. Then, if the liftbag dumps on the surface because there's not enough suspended weight to keep it upright, the CO_2 in the pop-buoy will have expanded enough to keep the scooter from sinking.

Another method is to strap squashable material to the scooter, such as neoprene. This will enable the scooter to float by itself on the surface, but as the cells crush with depth the buoyancy is reduced. In this case, if you send the scooter up on a liftbag, or release it from a decompression stop, or simply place it in the water prior to your entry, the scooter will float on its own.

A scooter with permanent positive buoyancy may be, under some circumstances, a liability, or at least a discomfort. For example, if you do a lot of decompression diving you might not want the scooter to exert a constant upward tug while you complete your stops. In that case you could clip a line reel to the scooter and let it go: it will float over your head while you remain at depth.

How you neutralize your scooter's positive or negative buoyancy depends upon personal preference, the type of dive, and day to day weather and water conditions.

Dig it

There's no doubt that scooters have a great many uses other than just towing a sightseer over a vast coral reef. Consider this event off the New Jersey coast.

On a wreck known at the time as the January Wreck, and later identified as the *Francis Perkins*, which sank in 1887, Gene Peterson fanned away some sand and found three gold coins. By the following weekend one local dive shop had sold three scooters.

The wreck was less than eighty feet long and lay in 80 fsw. So why the rush for auxiliary propulsion? Believe it or not, in wreck diving I've seen more scooters used as underwater earth movers than as towing units. For that reason I need to devote some space to an unusual type of employment for which the scooter was never intended, but in which capacity it excels.

Prop wash deflectors are an established and indispensable tool in shallow water salvage work. Once a surface support vessel is immobilized with a four-point moor, an angled shroud lowered over the propeller deflects the backwash downward. It's the ultimate in sand fanning. Likewise, a scooter whose propeller is aimed at the bottom from close range can waft away sand like an aquatic leaf-blower.

It's amazing how much bottom material a scooter can churn up and dispel. Held at an oblique angle over loose, granular sand it will dig a hole a foot deep in seconds. Clinging silt and heavy clay stir more slowly but can still be flushed out easier by means of spinning blades than with clam shells or Ping-Pong paddles.

You can always tell a scooter digger when you come across him on the bottom: he's surrounded by a cloud of dirt like the Peanuts character Pigpen. If you work the upcurrent side of a spot, the detritus will be carried away from you and visibility will not be reduced to total blackout.

According to Isaac Newton's third law of motion, every action produces an equal but opposite reaction. This means that while you're blasting sand out of a hole, you're being bowled across the wreck like a circus performer shot from a cannon. Since this is a natural law it cannot be repealed. But it can be amended.

The easiest thing to do is to brace yourself against wreckage and place the nose of the scooter against your chest. This works fine as long as you keep a good grip on the handles. If you lose control, the scooter will dart off under your arm or past your head, but won't go far because when you lose contact with the on/off button the propeller will instantly whine down.

If your scooter has a speed selector switch, try it first on low power. The major hazards of this kind of operation are the scooter nose banging you on the chin, or the propeller snatching loose gear such as mesh bags and gauge panels. Once you master the technique you can move a lot of sand in a short period of time.

Another workable method is tying the scooter to the wreck: in essence, establishing a one- or two-point moor so the prop wash hits the same

Uncovering a cannon. (Photo by Jozef Koppelman.)

spot continuously. This has the advantage that once in place there's no effort on the digger's part to push against a bucking machine. He can look for goodies in the downdraft. With a one-point moor you can push the machine from side to side in order to dig a wider hole or excavate a trench.

Do be careful of sand or silt clogging the switch. With all that fine-grained material flying about, some of it is bound to work its way into the slits around the throttle control and switch mechanism. Use your liftbag inflator periodically to blow out any build-up.

Another problem can occur when the propeller sucks in broken bits of shell, large-sized debris, kelp or seaweed, or loose lines or lanyards. Any of these foreign objects might jam the blades against the shroud or wrap around the shaft, thus forcing the motor to grind to a halt. If the motor is not switched off right away it could be damaged by overheating or it could burn up completely. If enough torque is exerted when the propeller jams, the shaft could be bent, a blade could snap off, or the shroud could be distorted. A further possible consequence of a bent shaft is leakage around the gland.

One way to prevent jams is to construct and

emplace a blade guard made out of chicken wire or heavy-duty screen. This is especially useful for preventing spare regulators from streaming into the propeller when using the scooter to scoot.

Perhaps someday someone will invent a two-part scooter for digging. The battery compartment will be connected to the propeller by means of a long flexible shaft such as that used in Dremel tools. This unit would be easier to handle than a one-piece scooter because the digger would have to hold only the propeller shroud, and because water could flow past the blades with less resistance.

Aquazepp - Monster DPV

The Aquazepp (and its derivatives) is the Harley-Davidson of scooters. At close to five feet long, it is a true diver propulsion vehicle that approaches the size of the Italian chariot that was its forebear. Two pairs of batteries give it lots of range and power to spare. The mini-zepp, its little brother, is one foot and two batteries shorter. It has the same voltage rating but its power output is reduced by half: that is, it goes just a fast but only half as far.

The Aquazepp was designed for clandestine military operations. According to the engineering specifications it can be taken to a depth of 300 fsw. Recreational scooters can't go as deep without special modifications: generally, internal ring structures like extra ribs that strengthen the casing so it won't deform under pressure and breach the o-ring seals.

The Aquazepp is less than admirably suited for deployment from a boat, or for circumnavigating shipwrecks. It's heavy, bulky, and awkward to handle topside. For protection, it should be stowed in a cradle that is secured to the deck or bulwarks: a contrivance that takes up valuable space where crowding is already customary. Launching the beast can be accomplished in only the calmest of seas, and then by means of a boom or by a bunch of strong-backed persons.

Although the Aquazepp glides like a charm underwater, the features that make it superior to its less demanding competitors are wasted in wreck reconnaissance. Range and endurance are attributes that are expedient only when they are needed, such as in the exploration of extensive cave systems. By comparison, wrecks are small and dives may be deep but short.

When the distance to be covered is great and the time spent in traveling is long, the Aquazepp is the undisputed master among DPV's that are commercially available.

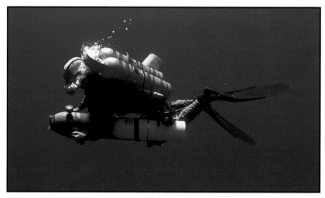

At left is an Aquazepp being deployed from a boat by means of block and tackle. Above is Billy Deans flying a mini-zepp. Note the compass mounted on top just behind the nose. The white cylindrical object behind Deans's elbow is a sling bottle, not the mini-zepp casing.

Cave divers have put these machines through some awesome paces, taking them on excursions to hell and back. Since road service isn't available in natural tunnels, a motorist in distress may kick frantically about his situation but still not make it out from where his DPV took him in. Therefore, they conduct these lengthy peregrinations in pairs or teams. That way, if one machine breaks down or runs out of juice, the pilot can be towed out by another DPV.

They have also converted Aquazepps from personnel carriers to tank wagons: staging platforms for back-up bottles and all kinds of accessory equipment. Homemade frameworks hose-clamped to the undercarriage have enabled cave

Paul DeLoach inspects a video camera mounted on an Aquazepp prior to an exploratory flight into Wakulla Spring. Notice the array of tanks slung under the belly. (Photo by Bill Stone, courtesy of Cis-Lunar.)

Hello, Dolly!

Billy Deans turned his mini-zepp into a camera dolly. In order for a movie or video camera to work properly on a DPV it must be securely fastened. Otherwise, the motion through the water causes the camera to flutter, much like a dog's tongue when it hangs its head out the window of a speeding vehicle.

Billy Deans and his roving video camera passing by the stern of the *Monitor*.

Deans employs a unique mounting bracket that permits him to turn his camera from side to side so he can speed along the hull of a wreck and capture it in its entirety, or keep the camera focused on a single spot as he pivots around a turn. He welded a handle onto the locking nut, thus giving him the leverage to loosen the camera whenever he wants, so he can change its direction of aim.

divers to haul a warehouse full of stage bottles deep into the Earth so close to the devil that the water began to boil.

During the 1987 Wakulla Spring project these super-DPV's were used to push the limits of exploration far beyond anything that had been done before. Because of the ungainly configuration presented by one Aquazepp that was heavily laden with cling-on tanks, it was appropriately dubbed the Klingon Kruiser.

The only rule of thumb is: if you can get it into the water you can attach it to an Aquazepp.

Wreck divers could use such powerful DPV's to reach shipwrecks that are located close to shore. This would save money on boat charter fees.

What's Next, Doc?

Diver propulsion vehicles may seem to have come a long way since the early days of diving, but they've actually only come full cycle: from a motorized explosive device operated by military frogmen to the high-speed cylinder transport wielded by Florida cave divers. It can get you where you want to go, carry your gear like an underwater Sherpa, and uncover treasures that might otherwise have gone unfound.

Not only that, but the reduction in exertion helps to prolong your air supply while diminishing the build-up of carbon dioxide.

Already in the works is a dual DPV system with parallel motors and twin screws, each with its own compartment: an underwater catamaran.

Like wonder, ingenuity never ceases.

SUBMERSIBLE AIR DECOMPRESSION COMPUTERS

I once wrote that the U.S. Navy Standard Air Decompression Table was the Bible of decompressing diving. This is still true. Both the Table and the Bible represent basic principles that at one time were all we had for guidance. Now, in an era of enlightenment and computerized revelation, their importance lies more in the realm of ancient history: quaint moralistic fables superseded by the modern mores of a complex and fast changing world. This is not to say that both do not have their place in today's society, only that their limitations must be recognized.

Let's clear away some misconceptions about the Tables.

Navy Tables for Navy Divers

The most significant feature about the Navy Tables is that they were developed for the Navy. That may sound overtly obvious unless you fully comprehend what it means. Let me quote Captain Claude Harvey of the U.S. Naval Submarine Medical Research Laboratory:

"One of the burdens the Navy has carried over the years is that we would develop tables for certain use by a certain Naval population, then when other people would adapt them for other uses sometimes they were not ideal, and then they would come back and say, 'Navy, you are producing horrible tables.' For our purposes they were fine."

Dr. Richard Vann of the Duke Medical Center commented on Captain Harvey's statement: "There has to be a change in the appreciation of divers . . . that there is no magic depth and time limit that is going to eliminate the risk of decompression sickness. It is a question of education."

And there's the rub. Because the National

Oceanic and Atmospheric Administration promotes the use of Navy Tables in the NOAA Diving Manual, and because most civilian instructional agencies taught Navy Tables almost exclusively, American divers have gotten the mistaken impression that those Tables were infallible and sacrosanct. Far be it.

Fortunately, the training agencies have figured this out and have now moved away from Navy Tables to more conservative Doppler tables. But the old guard dies hard.

Unacceptable Risk

With military applications in mind, Navy Tables were designed to get Navy divers out of the water as quickly as possible with an acceptable occurrence of decompression injury (DCI). Acceptable? In my mind, *no* occurrence of DCI is acceptable. But in the Navy the situation is different because Navy divers are not permitted to dive without an onboard recompression chamber, a fully accredited operator, and a doctor or medical technician: a luxury that civilian divers cannot afford. If a Navy diver gets bent after a dive he is simply shoved into the tank with a sandwich and a six-pack (of fruit juice) and cranked down until the symptoms of pain or paralysis are relieved. Then he is brought up slowly.

U.S. Navy Tables were developed to prevent the bends, not to prevent the formulation of bubbles.

Besides being a disservice to Naval personnel by ignoring the harmful long-term effects of constant injury by bubbles only slightly small enough to pass through the cells and bloodstream without causing symptomatic DCI (one effect being aseptic bone necrosis developing later in life), the incidence of DCI with Navy Tables is not emphasized by those agencies teaching the use of the Tables.

This is not the Navy's fault but it *is* your problem. You must be fully aware that inherent in the applicability of the Navy Tables are the procedural and mechanical back-ups provided by Naval administration (doctors and chambers). Does this scare you? I hope so. Here's something else that will scare you.

The current rate of DCI among Navy divers making decompression dives averages one percent (1%): a frightening statistic when translated to the recreational diving community. It means that the typical dive charter operator carrying fifteen to twenty divers a day would have at least one bends victim every week. The weekend dive boat would be calling for helicopter evacuation once a month. And each advanced or technical diver would take chamber rides on an annual basis.

Think about that the next time you rely solely on Navy Tables either to calculate your decompression or to go right to the edge of a no-decompression dive.

U.S. Navy Tables have the shortest decompression profiles in the world.

They will not only get you out of the water faster than any other table, they will get you to a chamber quicker.

Also missing from the equation is the fact that Navy divers seldom make repetitive dives. Instead, they'll stay down for an hour or two breathing through an umbilical with an unlimited air supply, decompress in the comfort and security of a chamber, then take the rest of the day off from extreme physical activity.

Cave divers learned the hard way that you can get bent from the exertion of climbing out of a steep-sided sink hole. How many wreck divers, I wonder, can attribute the onset of DCI to the rigorous post-dive strain of struggling up a ladder on a rocky boat, fighting a pitching deck, or helping to pull up a fouled grapnel?

Out with the Old

Assuming all this is true, why haven't we seen a higher incidence of DCI among avid technical divers, where decompression diving is com-

monplace? Partly because by the time a diver has reached an advanced stage in diving he has learned to be more conservative in his approach to decompression from those who have already paid the price of education.

The "old-timers" add fudge factors to the Tables by calculating their decompression requirements for the next deeper depth and for the next longer time increment, or by breathing oxygen at the 10-foot stop. Moreover, today's divers can rely upon sophisticated electronic devices that compute their decompression penalties more accurately and more conservatively than Navy Tables.

The advantage of Navy Tables has never been their reliability, but their availability. Even in their heyday they did not perform adequately for deep decompression diving. Not only was the square profile unrealistically restrictive, but the repetitive dive group letters stopped at 190 fsw - where many of today's technical dives begin.

Ingrained in the minds of the general diving populace is the unfounded belief that deep repetitive dives cannot be made. There is no physiological basis for this belief, only the observation that the Navy halted experimentation on repetitive deep diving at 190 fsw because deeper repetitive dives were not part of the Navy regimen. For "exceptional exposures" (200 feet and deeper) Navy divers make one long dive and call it a day.

A Stopgap Solution

To overcome this deficiency in Navy Tables, before the advent of reliable computers, I was forced to extrapolate my own repetitive group letters from the 190-fsw depth by extending mathematically the curve on which Navy Tables are based. For example, twenty minutes on the bottom at 210 fsw requires a total ascent time of 40:30 min:sec. The next largest total ascent time on the 190-fsw table is 44:10 min:sec, which is for a twenty-five minute bottom time. I plugged the repetitive group letter M into the residual nitrogen timetable for the *next shorter* surface interval than the one I actually had. So, if my surface interval was 3:09 hour:min, which yielded the new group designation E, I used F. In that case, the "bad" time (residual nitrogen time) for 190 fsw is eleven minutes. Then, using the next five-minute increment at the next deepest depth, I could make a fourteen minute repetitive dive to 210 fsw, and decompress according to the thirty minute table for 220 fsw. (Thirty minutes is the total of eleven minutes bad time, fourteen minutes proposed bottom time, and a five minute G-factor that I threw in for the heck of it.)

Granted there was some guesswork in making these calculations, and not a little trepidation in performance since my extrapolation was strictly theoretical, but the system worked because of the conservative nature of my formula and the fudge factors I incorporated. The proof is in the pudding: I didn't get bent making repetitive dives to depths over 200 fsw.

By contemporary standards my total ascent times for repetitive dives with exceptional exposure were overly conservative. Ironically, they correlate pretty closely with times that modern computers recommend.

Alternative Tables

My point in this discussion is not to ridicule Navy Tables, but to firmly establish in your mind that they were never intended to be used by recreational divers, tourist or technical, where constant mobility and multilevel exploration are the norm, and where surface support is at a minimum. Navy Tables are nice to fall back on when all else fails, as long as you incorporate some fudge factors as I explained above, but there are better alternatives available. In the computer revolution, Navy Tables are going the way of the slide rule.

If you are really hung up on the use of tables in defiance of modern technology, or like to carry a set as a back-up, I suggest you investigate those that are more conservative than the U.S. Navy's: DCIEM (Canada's Defence and Civil Institute of Environmental Medicine), RNPL (U.K.'s Royal Navy Physiological Laboratory), Buhlmann's Swiss tables (the printed form of the software program used in many dive computers), Huggins' tables (also used in dive computers), or any other Doppler table.

Compared to U.S. Navy Tables, all those listed in the previous paragraph have shorter no-decompression limits and longer decompression requirements for specified depths and times.

I Said, Out with the Old

Most divers are blissfully unaware that in 1976 NOAA published "Recommendations For Improved Air Decompression Schedules for Commercial Diving" in which researcher/author Edward Beckman "recommended that the British Royal Naval Physiological Laboratory air diving tables of 1968 be adapted for use by the commercial diving industry in this country in an effort to provide greater safety to diving decompression practice."

In the course of his studies, Beckman found that "the use of the U.S. Navy air diving tables for commercial diving had not only produced severe decompression sickness but also caused a high incidence of dysbaric osteonecrosis as well," and that "the uses of the *U.S. Navy Diving Manual* may not be suitable for use as a standard for commercial diving operations, at least insofar as diving tables are concerned." NOAA neglected to mention this study in its own diving manual, whose decompression tables are identical with Navy Tables.

In with the New

In technical diving, the aim is not to be the first one out of the water, but not to get bent. Even though your companions may rack up less hang time than you, they will not get back to the dock any sooner. You're all on the same boat.

The progressive diver must keep up with advancing technology not only so he can make his dives safer, but so he can continue to expand his horizons while becoming more effective during his time spent under water. Considering my penchant for making repetitive deep dives when such a practice was abhorred or, at best, frowned upon, I for one was overjoyed to see decompression computers come along.

Let's talk about their uses and abuses, types and functions, which features are indispensable for the serious technical diver, and which bells and whistles are worth tolling and blowing.

A Brief History - The SOS Decom Meter

Submersible decompression calculators are not new in diving. I bought one in the early 1970's (when it had already been on the market for a decade) that was disparagingly called the Bends-O-Matic by people who didn't trust it and, consequently, had never used it. I no longer have mine, but I still have the instruction pamphlet.

The Automatic Decompression Meter (or Automatic D.C.P.) was made in Italy by SOS. The wrist-worn unit consisted of a flexible bladder that forced air through a porous ceramic filter into a pressure-proof chamber where it actuated a bourdon tube. As the bladder was squeezed by water pressure during descent, air diffused through the ceramic filter into the chamber, where it expanded the bourdon tube and caused the attached needle to deflect clockwise around a calibrated faceplate. The deeper a diver went and the longer he was down, the more air diffused through the filter and the farther the needle moved. Upon ascent the air diffusion was reversed, thus reducing pressure on the bourdon tube and permitting the needle to reverse direction. Theoretically, this two-way diffusion rate simulated nitrogen absorption and elimination in the human body.

Although the decompression stages were scaled in 10-foot increments, the diver could follow the needle foot by foot during a decompression ascent. Because the stops were painted in red, in the lingo of the day "in the red" meant that the diver had entered the red zone and required decompression. Interpreting the gauge took a little getting used to because longer dives came out of the red on a different scale than shorter dives.

The weak link in the "decom meter," as we called it, was that the flow rate through the ceramic filter was predicated upon nitrogen absorption and elimination on an average of all body tissues, and closely approximated the Haldanian curve on which Navy Tables are based: a system I have already vouchsafed as inapplicable for technical diving purposes.

Furthermore, while the decom meter was more conservative than Navy Tables at depths less than 50 fsw, it tended to be less conservative at greater depths. Worse yet, the air exchange rate

caused the meter to eliminate the calculation of residual nitrogen after only six hours of surface interval. Simplicity begot some general inaccuracy.

More History - Buhlmann and the Deco-Brain

The big breakthrough for decompression computers came in the mid-1980's with the introduction of the Swiss made Deco-Brain II. (The original Deco-Brain that was introduced in 1983 incorporated a table-based program that considered only one tissue group, and ignored the others.) Using an ultrasonic Doppler to monitor micro-bubbles in the bloodstream, Dr. A.A. Buhlmann developed an algorithm for nitrogen absorption and elimination that predicted decompression requirements for all depths and durations to 330 fsw for 167 hours - nearly enough for today's technical diver.

The theory behind Buhlmann's algorithm was that instead of decompressing in such a way that the only bubbles allowed to form were those small enough to circulate harmlessly, *no* bubbles were allowed to form. These no-bubble tables furnished a completely different dive profile than Navy Tables. For the first dive, no-decompression limits were shorter, and decompression dives required longer decompression times; but repetitive dives might require less decompression when compared to Navy Tables.

The first time I used the Deco-Brain II it scared the heck out of me. I tested it on the wreck of the *San Diego*, an armored cruiser sunk in 110 fsw off the south shore of Long Island, New York. Because it rose 45 feet off the bottom (to 65 fsw), the fudge factor I employed with the Navy Tables was to use the 110-fsw maximum depth as per instructions, knowing that most of my bottom time would be spent at depths between 80 and 100 fsw.

On the first dive the Deco-Brain II held me down ten minutes longer than the Tables: hang time that seemed to be excessive and wasteful in comparison with my experience. But on the second dive it told me to come up fifteen minutes *earlier* than the Tables: a frightening prospect for one who yet lacked faith in the system.

My heart fluttered as I slowly broke the surface, kicked back to the boat, and climbed aboard. I sat fully dressed for half an hour, ready at the first twinge of pain or notice of numbness to jump back in and do a miss-stop decompression. My legs started itching with what I finally determined was a reaction to the material of my longjohns. Eventually, after feeling nothing but psychosomatic tingles that crept up my spine, I accepted the validity of electronic computer diving. A new age had arrived.

The Deco-Brain II had quite a few innovative features. As an electronic device it relied upon solid state circuitry, utilized light emitting diodes (LED) for data readout, displayed information numerically (instead of using needles like analog gauges), and flashed red warning lights for exceeding the ascent rate, for reaching each required decompression stop, and for going out of range of the unit's maximum prescribed depth and/or time.

It displayed total dive time, the present depth, the maximum depth thus far attained on the dive, the decompression ceiling, and the total decompression time required. As they say, it did everything but wash the windows. And because the unit was so heavy, it allowed me to take several pounds of lead off my weight belt. If I wore the unit on my wrist I was unable to lift my arm; strapped to my gauge panel, before long the weight split the high-pressure hose.

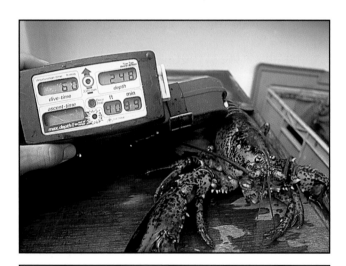

My Deco-Brain II after a dive on the *Andrea Doria* - and a lobster that I grabbed on the bottom.

In surface mode the Deco-Brain II scrolled no-decompression limits to a depth of 140 fsw (a function I have always found useless since, even today, no decompression computer scrolls to the depths where I plan to take it), flashed logbook values, indicated the minimum safe time until the user could fly on a commercial aircraft, and counted down the total desaturation time. In addition, the software package could be updated as refinements to the program were made.

Unfortunately, the plastic casing in which the hardware and batteries were housed developed micro-cracks after long immersion in sea water, thus flooding the unit and destroying it. It would do no good to replace the damaged components because the watertight integrity of the casing was permanently compromised. The unit became a throwaway.

When the going got tough, the manufacturer disclaimed all obligations to purchasers. Divetronic AG refused to honor warrantees and accepted no responsibility for poor workmanship. It would neither repair nor replace damaged units, preferring instead to let the customer bear the burden of loss while it went on to other money-making schemes. There's a lesson to be learned here about business ethics.

Years later I had a similar problem with Beuchat. I purchased an Aladin Pro directly from the company, but when I took it out of the box and used it for the first time, I discovered that the display was unreadable because some of the LED's were nonfunctional. Beuchat refused to repair or replace the unit, claiming that I must have caused the damage by dropping it. They offered to sell me another unit at cost (and at no loss to the company). Imagine if you bought a new car that was totally undriveable, and you were told by the dealer to junk it so he could sell you another one at half price!

Half-way Computers

After the Deco-Brain went the way of the dinosaur there was a hole in the field of decompression computers. To be sure, a plethora of companies entered the market with devices they euphemistically called "dive computers." Ostensibly, this nomenclature describes an instrument whose purpose is to track a dive to the no-decompression limit and, in an emergency, somewhat beyond. Most of them do this quite well.

This type of computer doesn't meet the needs of technical diving. The designation "dive computer" appears to be a marketing stratagem whose purpose is to contradict a more accurate description: "no-decompression computer."

Strangely, some computers that have sufficient microprocessor capability to compute decompression profiles are also merchandised as "dive computers," possibly for liability reasons or to cover inherent inadequacies. In most cases, these units lack some essential characteristics necessary to fulfill the roll of a true, full-fledged decompression computer.

For example, they might go out of range at a ridiculously shallow depth, say between 150 fsw and 200 fsw, leaving you to calculate your own decompression once you exceed the unit's limitations. This type of computer is designed for tourist diving only.

Thankfully, there are units on the market that fill the bill for technical diving.

Full Service In-water Decompression Computers

Before we go on I suppose I should offer an explanation for the seemingly cumbersome chapter title, and the blatantly awkward section heading.

A few years ago the distinction between decompression computer types wasn't necessary. The only ones on the market were designed to be worn on the wrist (although a couple of no-decompression computers could be secured to the high-pressure hose). The "wrist-worn decompression computer" was shortened to "decompression computer," or simply "computer." In the technical diving venue and vernacular, "computer" was sufficiently descriptive.

Despite the ergonomically curved back wall of the casing, I strapped or cable-tied my computers to my gauge panel. I don't like encumbrances on my arm. After this sartorial method became popular, some manufacturers altered the dimensions of their units so they could fit inside of gauge panels. They then went on to design integrated gauge panels in which the decompression computer was the central device, accompanied by the usual array of ancillary gauges.

At this point it became somewhat imprecise to think of all decompression computers as wrist-worn units. Yet all was well as long as "computer" was understood to be a generic term for all in-water models whose differences were slight. Now, with the advent of personal computer decompression programs (see "Mixed Gas Helium") and the proliferation of laptops that are brought to the dive site, "computer" is no longer a morpheme and its usage is confusing.

We need discrimination in terminology. "In-water" computer suffices to differentiate the watertight version from its topside counterpart, the laptop. But I prefer "submersible." It worked when pressure gauges were first taken into the underwater realm; they became submersible pressure gauges, or SPG's.

I don't intend to introduce an acronym such as SDC (submersible decompression computer), but I do propose to characterize computers by their essential distinguishing feature: to wit, whether it can get wet or has to stay dry. "Submersible decompression computer" can be the formal term, and "wet computer" the slang.

This implies that PC's and laptops be known as "dry computers," but I think that "laptop" is already ingrained in the language of society and is not likely to change.

Plan Ahead, but Dive a Framework

Technical diving is a discipline that requires the generation of data in a comprehensible format that is fluently displayed, so the diver can monitor the constantly changing status of his dive. Since dives to the frontier tend to be exploratory rather than planned, you must leave room for variation and have the latitude to deal with circumstances as you find them.

For example, suppose you're diving a wreck that no one has ever been on before. You know nothing about the layout of the hull, how large it is, how high it stands, or how intact it may be. From the boat you can't perceive how well the grapnel is hooked, and there's no way to predict what the current and visibility will be on the bottom - particulars that will affect the conduct of your dive.

Against this array of unknowns it's not reasonable to assume that you can "plan your dive and dive your plan" with any amount of accuracy. The only parameters you can define with relative certainty are the approximate maximum depth and how long you'd like to stay down - and these are subject to change.

This doesn't mean that you should jump into the water without any idea of what you're going to do or how long you'd like to spend doing it. It means merely that your outline should be flexible. Plan a framework around which contingencies are built and within which opportunities can be grasped.

So, Full-Service Wet Computers

Personal preference carries a weight all its own in the selection of merchandise. Consequently, in order to satisfy the demands and desires of the total diving community, computers (abbreviated here because that's what this section is all about) offer a wide variety of forms, features, and functions. By definition, however, a full-service computer has everything - and then some.

The greatest benefit derived from a wet computer is complete and total knowledge of your decompression status at all times. In combination with air endurance predictions, this enables you to make decisions *during the dive* about how long to continue it or whether to terminate it.

If you're fifteen minutes into a 200-fsw dive, what kind of information do you need to make an intelligent decision about whether to spend time looking for the grapnel, tie off a gap reel and check out that side passage, or head for the surface? Two items: remaining air supply and total decompression penalty.

Notice that I wrote *total* decompression penalty. It does little good to know only that your ceiling is 30 fsw and that the length of that particular stop is five minutes.

Informed Decision

By way of example, let's assume that you knew you were facing forty-five minutes of decompression. If you had 1,000 psi remaining in twin tanks, chances are that you could last out the hang with no problem. If you had 2,000 psi, you might want to stay down a little longer. If you had 500 psi you could be cutting it close. But in order to make any decision other than the default (to ascend), you must have the forty-five minute figure in hand (or on display).

Let's look at it another way. As a general rule of thumb, with large capacity tanks and with proper air management, most of the dives I make to intermediate depths are not restricted by air supply but by the amount of decompression I'm willing to endure. Assuming optimum conditions, I might consent to hang in total boredom for a couple of hours. But if the water is cold, if the current is strong, if the waves are high and choppy, or if I'm feeling fatigued, I might reconsider how much punishment I can withstand.

But suppose the conditions improve or worsen partway through the dive? In either case, I want the capability of modifying my original game plan. If I feel good in the water, if I'm not working hard, if I'm warm and comfortable, if I'm staying shallow and saving air, and if there are no adverse surface conditions, I'll extend my bottom time accordingly. On the other hand, if the water is colder than I expected, I'll cut the dive short so I don't shake to pieces during the deco.

The point is that in order to make these judgment calls I need ongoing factual updates so I can re-assess the circumstances as they occur and alter the conduct of the dive accordingly. I watch my computer tack on the minutes of required hang time, make sure that I have sufficient air to complete the predicted deco, and plan my ascent accordingly.

To make decisions like those described above, you don't need a degree in hyperbaric medicine, you don't have to understand the theory of complement proteins, you don't have to know M values or tissue half-times or hemostatic mechanics, and you don't have to be a doctor or a physiologist or a certified scuba instructor. What you need is experience in the water, confidence in your equipment, and familiarity with technical diving practices.

Plus a decompression computer that can handle the job.

A Conservative Approach

Of course, it's nice to have faith in the computational model on which your computer program is based. But even the simplest "dive" computer on the market is safer and more conservative than Navy Tables. By more conservative I mean that the no-decompression limits are shorter, actual decompression requirements are longer, and the ascent rate is slower.

In my opinion, more divers get bent by ascending too fast than by not decompressing long enough.

I am a staunch advocate of the 30-feet-per-minute ascent rates prescribed by most decompression computers. This slow rate of ascent is built into the algorithm, and is primarily responsible for the prevention of bubble formation. Some modern computers offer variable rates of ascent: faster at deeper depths where the pressure change is small, slower near the surface where the pressure change is the greatest.

A slow rate of ascent is also partially responsible for giving longer bottom times on repetitive dives: without preformed bubbles from the first dive, you don't have to worry about them growing larger on successive dives.

The other reasons for longer repetitive dives are allowance for multilevel excursions and credit for every second of surface interval (instead of the squared-off surface intervals that are built into Navy repetitive tables).

Furthermore, all the calculations are done for you on a continuing basis. There's no need to refigure your decompression penalty when you overstay your time by a minute or two or reach a slightly deeper depth.

Individual Tolerance to DCI

Naturally, no decompression model is perfect - nor can a mathematical construct ever be perfect. Just as some people are more susceptible than others to colds and flu, so are some divers more susceptible to the bends. I have friends who won't switch to decompression computers because their bodies tolerate Navy Tables without a hint of trouble. They see no need to incur what they perceive to be unnecessarily long decompression times.

Personally, when I used to dive Navy Tables to the edge (that is, when I came up exactly when they told me to) I often found myself feeling fatigue and having headaches, and sometimes felt subclinical symptoms of DCI such as muscle aches, tingling, and an increased tendency of my skin and extremities to "go to sleep" if I sat in the same position for too long. My body was telling me that the decompression performed was insufficient.

What possible long-term effects these conditions might have on the human body have yet to be determined.

Transitory Susceptibility to DCI

Besides individual tolerances to DCI, susceptibility can be affected by fixed physiological distinctions as well as temporary physical states and chemical imbalances. Of the first category is hull configuration: obesity, circulatory efficiency, past injuries and scar tissue, and overall fitness. Of the second category is amount of sleep, hard work, alcoholic content, seasickness (or any sickness), medication, and dehydration.

There was a case of a woman who was so afraid of urinating in her wetsuit that she severely reduced her fluid intake for several days prior to a dive trip. She got bent as a result. I'll often forego that second cup of coffee in the morning before a dive, but let's not go overboard with it.

Many people believe that cold water increases the risk of DCI. Strictly speaking, this is not true. Immersion in a cold medium has only a peripheral effect - like wind-chill when you're sitting inside a car. *Feeling* cold and consequently altering the blood's circulatory pattern, *will* affect sensitivity to DCI. A diver adequately dressed for conditions (such as one in a drysuit and thick layers of underwear) and who doesn't feel cold, has little to worry about on that score.

Computer Caveats

So, how do computer programs take all these variables into consideration? Generally speaking, the same as Navy Tables: they don't.

It is necessary for you to understand that, like the SOS decom meter, the decompression computer tracks an approximation of the body's rate of nitrogen absorption and elimination. The computer does it more accurately by taking into account the half-life of up to sixteen tissues (sometimes called compartments). It factors in the safeguard of a slower ascent rate. And the resultant algorithm has been tested on human subjects. Furthermore, where some programs have been found through experiment to be unfavorable, the programmers have fudged the algorithm in order to make it more conservative.

Is this enough? Let's put it this way: is a speed limit of 55 mph enough to keep highway fatalities down to a tolerable number? If not, it follows that if the speed limit were reduced to 45 mph there would be fewer accidents. Reduce it to 35 mph and there would be fewer still. Eventually, traffic accidents could be reduced to zero; but then no one would be going anywhere.

Eventually you just have to say: go with it.

Divers do get bent using decompression computers. That is the risk divers take when they enter the water. The only way to ensure yourself of not getting bent is to never go diving.

By the same token, if you never crossed a street in your life you would never get hit by a car. But in that case you would have to stay at home all the time. The risk of intellectual starvation from television onslaught might be a fate worse than a death that was quick and sure.

The choice is yours and yours alone. So are the consequences. Don't try to blame someone else - or an equipment manufacturer - for the shortcomings of your bodily make-up. That would be like denouncing your parents for not bequeathing you a healthier hereditary background. You must wear the genes you were given.

However, you can increase the odds in your favor. Keep in mind that there is an essential difference between computers and human beings - between mechanical systems and biological systems.

Computers make calculations,
people make decisions.

At no time does a decompression computer factor into its equations how cold you feel, how hard you are working, how much sleep you've had, or your level of hydration. These are things that only you can do. And this is where a protoplasmic brain excels over an electronic brain.

Enough theory. Now let's talk about computers in practice.

Computer Power

This section heading has a double meaning. In its broadest sense, power is defined as "the capacity to perform." As a unit of electricity, power is "a quantity of energy."

In addition to their primary function of providing information about the decompression status of a dive, computers have been imbued with a variety of other functions. Some of these functions are invaluable, some are a matter of personal preference, and some are of doubtful benefit. In the latter grouping are consumer items that either help to sell the product or that appeal to the recreational mindset.

Let's generalize these concepts in relation to computers, then discuss each one in detail.

Options of Dubious Value

Let's start out by striking a middle ground. A computer with a thermometer is handy for making log entries, but it won't affect your dive. You'll know whether or not you're cold; you don't need a gauge to inform you. Nevertheless, I happen to like keeping track of bottom temperatures as the season progresses, and noting variations from the norm.

On the other hand, I find graphic displays that register pixels or bar graphs to be distracting and a waste of battery power. Some people may relish beholding the pixels disappear or watching the tissue bars shorten, but those functions only duplicate information already given in a more concise and easier-to-read digital readout. This form of nonproductive energy consumption I call computer tuberculosis, or computuberculosis.

Totally meaningless for technical diving applications is the length of time you can stay at your present depth on the air remaining in your tanks. This is a nice feature if you're diving a shallow reef and want to know when to turn around and head back for the boat. But realizing that I have enough air to stay at 190 fsw for an additional eight minutes doesn't help me plan my decompression. This feature is useful only if the computer correlates remaining air with decompression penalty.

A computer that goes out of range at recreational depths will not meet the requirements of today's technical diver, or of today's recreational diver who would like the choice of extending his range.

Dumping an accumulated log into a company mainframe is not worth the annual fee. Very little benefit accrues to the user. But a personal computer interface is an exciting and valuable tool. I will reserve this function for a later section.

Computer Turn-ons

The majority of computers are narcissistic in that they turn themselves on when they hit the water. This is a nice feature to have if you're forgetful or lazy. I have trusted automatic switching too much, sometimes with results that were frustrating at best. In the main, decompression computers are so dependable that the user is lulled into a sense of unremitting confidence.

There is no place for faith in technical diving.

The problem with such absolute reliance is that you never know until you're on the way to the bottom whether the computer is working or not. By then it's too late to do anything about it conveniently. Either you return to the surface and correct the situation, or you default to your back-up. Both options are less than desirable.

What prevents a computer from switching on automatically? A dead battery for one. But more likely is too quick a descent.

A computer runs through a self-diagnosis mode when it first becomes conscious of its surroundings. That's the meaning behind the display of crazy eights. At the same time, the depth sensor adjusts itself to atmospheric pressure in order to establish a set point. It can do this in a couple of feet of water after the contacts have been shorted, but taking the unit too deep too fast upsets the delicate pressure balance before the sensor can settle out.

The computer has gotten lost in depth because it didn't have time to perceive the altitude of origin. Or it assumed that its depth sensor has failed because it is sensing depth when it thinks it's on the surface. It then shouts for help by triggering the error mode. Once this happens, the unit must be removed from depth and then completely dried before it can run the program anew. This means getting out of the water and desiccating the contacts with a dry towel or cloth.

An ounce of prevention, as they say. Switch on your computer before taking the giant stride.

Light up your Life

The same illuminating caveats I made about digital timers and depth gauges ("Gear in Depth") apply here. Under low light conditions, during the hours of darkness, and deep inside a cave or wreck, computers are hard to read. Hardest of all to read are dot matrix displays; stay away from them. And be wary of short-sighted computers with miniaturized displays. You want numerals that are large and clearly decipherable, not ones that make you squint.

At least one manufacturer has gone a step beyond LCD's and LED's, and released a model whose faceplate is backlighted by fiber optics. To conserve power, the lighting system can be switched on prior to the dive and switched off immediately afterward (although it cannot be switched on or off under water). One model has a light that is activated for five seconds whenever the display window is tapped.

Assault on a Battery

Energy conservation may not seem like an important issue when such small power requirements are at stake. But if a battery goes dead in the middle of a dive, your decompression could be seriously compromised. Fortunately, most computers have a mode which signifies that the battery is about to fail - and they are more friendly and reliable at this prognostication than HAL.

However, there are questions that you want to ask before purchasing a particular model. Forewarned is forearmed, as they say.

You probably won't notice the "low bat" symbol at home, where you can take preventive measures by installing a new battery. It generally comes to your attention during the pre-dive check or under water. By that time the dive is already underway. It would be comforting to know that you can finish the day without the computer crapping out and leaving you hanging, so to speak.

Some models offer the warning about one hundred hours in advance. That's a long enough lead time to complete most multiday trips. Then you can change the battery at a more convenient time.

Perhaps as important as lead time is whether the battery can be changed by the user. Different models run the whole gamut of alternatives, from a unit in which the new battery can be installed before the dead battery is disconnected (thus saving the information needed to make repetitive dives), to the unit that must be returned to the factory in order to have the sealed compartment opened and professionally resealed.

In between are units with other solutions to the problem. In one model the bad battery must be unsoldered and the new one soldered in place, requiring the user to own or have access to a soldering iron, and to have the necessary skill to use it. If a poorly soldered joint comes loose during a dive, you're out of luck with no one to blame but yourself. And you have to be careful not to overheat the battery in the soldering process or it can be damaged.

Some models have batteries that are as easy to change as flashlight batteries. Simply unscrew a plug that is o-ring sealed, let the battery fall out, insert a new one, and screw down the plug. Then there are computers whose battery compartment is separated from the factory-sealed electronics compartment by a watertight bulkhead. This way, if the battery compartment leaks it doesn't destroy the expensive computer circuits. All you lose is a battery.

Wrist Restriction

When I was a novice diver and wore my gauges on my wrist, I suffered a case of severe localized bends which not only paralyzed my arm and turned it purple from shoulder to fingertips, but caused me to pass out from the pain. It happened like this.

Before entering the water I tightened the straps of my gauges and decom meter so they wouldn't flop around the sleeves of my wetsuit. On the bottom, due to the ambient pressure at 140 fsw, the neoprene was compressed enough that the gauges became loose anyway. I shoved them farther up my arm where the muscle is thicker and where they stayed in place.

Upon ascent the neoprene expanded and, since it couldn't spread outward, it pressed against my arm where it pinched off the flow of blood and lymph. The pain was excruciating. On the boat I had to be helped out of my wetsuit jacket, at which time it became visibly apparent that something was wrong with my arm, although no one knew exactly what. (This was the early 1970's, when divers understood very little about DCI.)

I passed out. When I regained consciousness I was examined by a medical doctor who fortuitously happened to be on board. (This was Gene Della Badia, a psychiatrist.) My arm did not appear to be any kind of decompression anomaly that anyone had ever heard of, but because of the paralysis and incredible swelling he recommended that I try to alleviate the condition immediately by means of in-water recompression.

I agreed, and was helped back into my suit and plunked overboard with a spare tank. Della Badia went down with me and stayed with me for an hour, making sure that I remained awake and alert. (Contrary to popular belief, this was the only time I have ever been treated by a psychiatrist.)

I felt relief at 20 fsw. I stayed there until I regained flexibility in my fingers. Then I moved up to 10 fsw, where I stayed until I drained the tank. By that time I could make a fist and bend my arm. Afterward my arm was still pretty swollen, and it ached with a continuous dull throb, but at least the color had returned to my skin.

That night, due to increasing pain, I drove to the University of Pennsylvania and checked myself in for hyperbaric treatment. After two and a half hours in the chamber, and feeling noticeable relief, the hyperbaric doctor diagnosed my condition as lymphatic edema - swelling of the lymph system - only the third such case in medical history. He was so excited about my condition that he took a series of photographs of my still swollen arm for the medical texts.

For three days I had to have my arm held over my head in a sling, so that gravity could assist in the relief of pressure. I recovered fully within a week or two. Never again did I wear another gauge on my wrist. I don't have to point out the obvious moral of "wristriction."

Console your Computer

Billy Campbell has devised an ingeniously simple solution for my problem: a curved console made from PVC pipe. He glued a short section of pipe to the inside of his drysuit sleeve between the wrist and elbow. It's fat enough to slide his hand through, strong enough to strap his gauges on to, and does not compress or distort at any depth. His arm never gets pinched and his gauges do not come loose.

An extension of that idea is the external model. This requires a PVC pipe that is large enough in diameter so the inside curve fits snug against the outside of the sleeve. Wetsuits and drysuits may need different sizes. Cut the pipe lengthwise to the desired width, say a third or a half of the circumference. Drill holes near the corners for surgical tubing, which will hold the console in place. Drill holes on the face of the console where needed to strap or cable-tie gauges and computers.

A similar contrivance but without the gauges makes a dandy slate on which to write decompression schedules. It's called a wrist slate.

That Uncertain Ceiling

Rehashing and emphasizing what I wrote a few sections back, what features are essential for decompression computers to have? The single most important piece of information that should be displayed continuously is the total ascent time required before you can break the surface without fear of getting bent. (Ascent time refers to decompression penalty plus time of ascent.) Only by watching the minutes mount up can you make an informed decision about how soon to leave the bottom.

Some computers show only the ceiling, not total ascent time. Certainly, a readout showing that your first stop is at 50 fsw gives a strong indication that you're in for a long hang, but the information displayed is not as persuasive as a readout showing sixty-six minutes to surface. Ceilings alone merely offer an approximation that keeps you guessing throughout the dive as well as throughout the decompression.

I once made a twenty minute dive to 250 fsw wearing a computer that displayed the ceilings only. When the first stop given was 80 fsw I knew I was facing a long one. As with Navy Tables, each successively shallower stage required more stop time than the previous, until I reached the 10-foot stop where the ceiling readout flashed for so long that I was sure the computer had hung up. It was like decompressing in a timeless limbo. The computer took forty-five minutes to clear the 10-foot stop, during which time I didn't know if the computer was functioning correctly. Nor did I have enough information to update my status and plan for contingencies such as short air supply.

Even with a model that displayed total ascent time I had a similar experience. Diving the *Monitor* at 230 feet, I watched the hang-time minutes climb into the seventies, eighties, and nineties. When I left the bottom after twenty-five minutes (the wreck has only fifteen feet of relief) my first stop was ninety feet, and the total ascent time displayed was ninety-nine minutes.

Again I thought the computer had hung up because, although the ceiling depths decreased, the time-to-surface display remained on ninety-nine - and stayed that way for forty minutes. Why? Because the readout window had room for only two digits. The computer calculated my decompression penalty accurately, but it had no way to display all the data until the time dropped down to the double digits. Not until the remaining hang time was reduced to ninety-eight minutes did I realize the extent of the decompression awaiting me. After one hundred sixty-five minutes I came out of the water with skin like a rotting prune.

The Fallacy of Total Ascent Time

Usually left unexplained in computer instruction manuals is the consistent inaccuracy of the total ascent time readout. Since time to surface is calculated on a "pure" decompression - made at the precise rate of ascent and stopping at the exact depth on which the mathematical model is based - no account is taken for variations on the theme.

Extra time is added if you ascend too slowly and if you stop at a depth that is deeper than the one prescribed. It's nearly impossible to maintain a faithful rate of ascent, especially climbing up an anchor line whose angle is difficult to judge, and while contending with a current.

Maintaining a constant decompression depth is feasible in most caves and springs where you can find a piece of bottom whose depth coincides with that of your ceiling. But in the open ocean this is practically impossible.

- If you rise as little as one foot above your ceiling, the computer either stops counting down your decompression time or it adds a second and a half for each second you remain above your ceiling. If you stay there too long, it locks up and leaves you hanging.
- If you sink as little as one foot below your ceiling, the computer adds on time because of the increased depth.

If you compare the second sweep of a watch with the computer's countdown display, you will observe that the countdown minutes last longer than sixty seconds. Thus when you settle in to do your decompression, you'll find that the time you spend clearing the computer exceeds the total predicted time that was originally displayed.

Furthermore, although a computer supposedly takes ascent time into account when calculating the total time required to decompress - and displays that figure on the bottom - on deep dives this figure is always underestimated. You might ascend for a full five minutes, reach your first stop, then discover that your total deco time hasn't been reduced at all. Presumably this occurs because your body is still absorbing nitrogen during the deeper portions of the ascent, so the computer tacks on this unanticipated nitrogen loading.

The total ascent time predicted by a computer reflects a perfect rate of ascent and exact depth stops which are impossible to achieve in actual practice. Any deviation from the ideal ascent profile results in additional decompression penalty. By any definition or stretch of the imagination, this means that on the average deep decompression dive the predicted ascent time falls short of the actual ascent time by as much as ten or twenty percent, possibly longer.

This is a quirk in computer calculation that you need to understand, so you can make appropriate decisions that are consistent with reality.

Keeping Abreast of your Computer

To ensure an accurate calculation of the absorption and elimination of nitrogen, you should keep your computer at the same depth as the midpoint of your chest. As you study the readout and act accordingly, remember that what you are doing in effect is decompressing your computer.

Rate of Ascent

It goes without saying that you need to know how deep you are as well as how deep you've been: the latter so you can do rough calculations in your head in order to check up on your computer; the former so you can stay in touch with your present situation, can monitor your ascent rate, and know when to stop.

An ascent rate of 30-feet per minute is a nudibranch's pace. It's as hard to hold down your speed of ascent as it is to drive 15 mph through a school zone. For that reason, a warning mode for exceeding the ascent rate is invaluable.

My Deco-Brain II had a red flashing light for this, but that feature has not been picked up by other manufacturers, probably because annunciator bulbs drain power. I was able to make sensitivity-training ascents on night dives without ever turning on my dive light.

A good computer usually displays this information with a flashing display. Some models sound a sonic alarm should a diver be so careless as to double the proper rate of ascent.

A Change of Pace

Modern computers also vary the rate of ascent, allowing the diver to ascend faster during the deeper portion of ascent (where the change in pressure is small) then slowing the pace upon approach to the surface (where the pressure differential is greater). This sliding scale optimizes decompression.

Older or less sophisticated models that maintain a contant slow rate of ascent are less efficient. This is because at the deeper portions of the ascent more nitrogen is absorbed than is eliminated. The net effect is one of increasing decompression penalty rather than reducing it.

If you think it's difficult to crawl up the water column at 30 feet per minute, wait until you have to slow down from 60 feet per minute to as slow as 20. The change of pace is nerve wracking.

Desaturation Time

Another built-in safety factor that computer algorithms have is an extended total desaturation time: that is, the theoretical time it takes for your body to clear itself of all residual nitrogen. Whereas the SOS decom meter desaturated after six hours, and Navy Tables after twelve, modern computer programs may not desaturate for more than twenty-four hours - depending upon the depths and durations of your most recent dives.

This method more closely follows the actual rate of elimination of nitrogen from the slowest tissues. Thus you are penalized for multiday diving because on any subsequent day you haven't completely desaturated from the day before. This is another safety feature offered by computers but ignored by tables.

The time remaining for total desaturation is displayed when the computer switches automatically to surface interval mode. The figure counts down for ease in tracking.

Flight or Fright?

A correlative of total desaturation time is safe time-to-fly: the amount of time you must wait before your body has eliminated enough nitrogen so you won't get bent at altitude in the reduced cabin pressure of commercial jet aircraft (which could be equivalent to altitudes as "high" as 8,000 feet).

I once got off a dive boat after making two long decompression dives to 140 fsw, got into an unpressurized private plane, and left the island for a five hundred mile flight. Even though the pilot hadn't dived that day, he had been on the boat so I thought he would fly the plane acceptably low. From my position in the back seat I couldn't see the instrument panel.

After a while I noticed an itching in my elbows and knees, and numbness in my fingers and toes. When I asked at what altitude we were flying he told me 10,000 feet. I asked him to descend to less than 5,000. The fuel economy was reduced in the denser atmosphere, but I felt a whole lot better as the itching and numbness went away.

In surface interval mode, computers with a time-to-fly option will automatically display the amount of time that must pass before it is safe to ascend to an appreciable altitude. The countdown is trackable the same as desaturation time.

Don't fly until the time has zeroed out. And don't climb tall mountains right after a deep dive.

Altitude Compensation

In a similar vein, a full-function computer should have altitude compensation. I mention this because today's technical divers are reaching higher as well as deeper by diving in mountain lakes and caves, or in the waters of high plateaus. For this you need a computer that isn't afraid of heights.

Some models can't be used higher than 1,000 feet above sea level. At the opposite extreme are models that adjust automatically to altitude as high as 13,000 feet, as long as they're turned on first in air. You'll need high altitude capability if you ever decide to search for Noah's *Ark* on top of Mt. Ararat.

PC Interface

Some models now feature a data log that can be uploaded onto your computer at home, stored permanently on diskette or hard drive, displayed on the big screen for easy viewing in color, and printed out in hard copy. This feature can also be incorporated with a logbook data base that allows you to enter additional information about the dive. With the arrival of the electronic logbook, the book and paper log will soon go the way of the dodo. Furthermore, the electrons are 100% recyclable.

It works like this. The wet computer manufacturer offers a software package that is sold separately and that is configured on your personal computer. At home, a cable links the two computers so the raw data from the wet computer can be downloaded through the interface. Cumulative data is then stored on the PC and displayed on its monitor. And not just numerical or statistical information, but a full graphic display of each and every dive.

For those with a scientific interest, a permanent record is kept of tissue half-times, nitrogen levels, oxygen toxicity units, and so on.

An entire dive profile from beginning to end can be reproduced on the screen. The display format is similar to that of a stock market report that shows annual activity and price fluctuations. But instead of projecting the ups and downs of profit swings, the graph shows descents and ascents (in the vertical vector) measured against the passage of time (across the horizontal vector). You can review a dive at a glance as if you were looking at a picture of it.

This is even better than reconstructing a dive by listening to a recording made by a communication system, because the display is visual. Vertical fluctuations are underwater movements through intermediate depths. The screen shows you how long you spent at each stage of the dive, both exploratory and decompression. Now you can go over a computer memory profile and correlate it with the dive in your biological memory.

Also displayed are maximum depth, decompression stops, and total ascent to the surface. Additional overlays allow you to note items of interest such as rates of descent and ascent, temperature gradients, decompression ceilings, and, if you have an air integrated model, air consumption during the various phases of the dive. Programmable wet computers will track the CNS clock and show gas switches.

Hoseless Tank Pressure Gauge

For people who don't like all those hoses hanging around, and who prefer to wear their computers and gauges on their wrists, hoseless technology is the way to go. Instead of screwing a pressure gauge hose into your regulator's high-pressure port, you screw in a submersible instrument package which reads the tank pressure electronically by means of a pressure sensor, then transmits the data to a wrist unit. Tank pressure is shown digitally and is integrated with the computer display.

Depending upon the model, the telemetry is accomplished in one of two ways.

In one method, the unit mounted on the regulator is a combination computer and sending unit. It contains the microprocessor chips that register the tank pressure, take the depth readings, keep the running time, make the decompression calculations, and so on. All this information is transmitted to a wrist-worn receiver and display unit - in essence, a drone with no computational functions of its own. It merely mirrors the data that is generated within the instrument package on your back.

The downside to this system is that any mal-

function in transmission - either in sending or in receiving - results in *no* data reaching the wrist unit for display. You lose your entire decompression profile as well as the depth, running time, and other information.

In the other method, the unit mounted on the regulator registers only the tank pressure, then transmits that information to the wrist unit for display. The wrist unit is a fully functional, self-contained computer. A transmission failure results in the loss of *only* the tank pressure reading. Your depth, running time, and decompression profile are not compromised in any way.

Heads-up Display

For years science fiction movie cyborgs and real-life aircraft pilots have been reading instrument data from heads-up displays. Now HUD is getting wet.

The underwater version of heads-up display utilizes the same technology as that employed in the Stealth bomber - without the stealth, bombs, and price tag. The backbone of the system is the hoseless, air-integrated computer described in the previous section. With a transmitter mounted on the regulator's first stage and a receiver strapped to the wrist, the operational network is already in place, and those who previously purchased such a unit are ahead of the game.

To achieve heads-up display capability requires the addition of a second receiver and an optical projection system, both tiny enough to fit inside a mask. These miniaturized components are sealed so they won't be damaged if the mask is flooded. The electronics are powered by small but powerful batteries. The mask is specially designed to house the batteries, optics, and electronics without being too heavy or bulky.

The receiver in the mask receives the same information as the receiver on the wrist. Instead of displaying the data on a panel, it shunts the display through a projector mounted inside the top of the mask. From there a reversed image is projected onto a prism located near the bottom of the mask. That image is reflected and reversed so it appears in normal reading mode outside the mask.

Ordinarily a diver peers straight ahead, in which case the virtual image is not visible. But if the diver looks down, the virtual image will appear to be displayed in the water at nearly arm's length below the mask - about where he'd be looking if he bent his arm to read the panel on the wrist unit (which displays the same information simultaneously).

The apparent size of the virtual image is slightly larger than the display on the wrist panel, and every bit as clear and sharp and bright, although the image is ghostlike in that you can see through it. A push button on the outside top of the mask permits the user to view any alternate display that is flashed periodically on the wrist unit.

Is the heads-up display a yuppie toy, space-age kitsch, or an indispensable tool? I vote for the latter. The clarity of the virtual image is not dependent upon visibility because it is generated from within the mask and only *appears* to be outside. In the darkest of nights, in the worst of silt-outs, the virtual image is bright and clear. No more holding your wrist unit to your mask, shining a light against the lens, and washing out the display with glare.

What happens if you and your buddy exchange wrist units? When you are within about five feet of each other, you can read your own data on the heads-up display and read your buddy's on your wrist. Both of you can track each other's decompression data. If you're at 200 fsw and can't find your buddy, and his gauge is showing 195 fsw, roll over and look at the blind spot above your back and there he is!

Can't take your hands off the anchor line because of a roaring current? You've got a massive camera rig that takes two hands to manage, and your console is behind your back? You're scootering through a long restriction and can't afford to let go? With a twitch of the eye you can read your depth, time, and deco information.

As noted in the previous section, a transmitter failure results in the loss of the display.

An observation: I understand the plural for hands-on, but heads-up makes me think of extraterrestrials.

Computer Care

Take care of your computer and your computer will take care of you.

Although most computers are rugged, well-built, and solid-state, don't treat yours like a pair of fins and toss it indiscriminately into a gear bag. Handle it gently with tender loving care like the sensitive instrument that it is. Pack it in a safe

place where it will not get knocked against other hardware.

• Don't leave your computer in the sun between dives; heat may cause the LCD's to fade and, if the casing gets hot enough, the plastic face might bubble or deform.

• If your computer is attached to your console, make sure tanks and other heavy items can't fall against it and crush it.

• When donning your tanks, don't let your console swing around and smash your computer onto rocks or against the boat's side.

• Hold your console in your hand, press it against your chest, or tuck it into your lap as you roll or jump into the water. If you're diving from a boat, this will prevent the gauge panel from hanging up on the rail and ripping out your high pressure hose.

• Mount a wire frame guard or clear plastic shield over the lens to prevent the facing from getting scratched.

I have my computers mounted on a console which is clipped to a harness D-ring by means of a length of surgical tubing, which is elastic. That way the console stays snug against my chest, but I can push it away in order to read the display.

Console Care

Because consoles are molded plastic and could possibly snap in two if accidentally bumped too hard, or can be blown off if the tank pressure gauge parts from its swivel seat, you could lose all the gauges attached to it. So here's a little secret that I learned from Jon Hulburt.

Drill a small hole in the plastic close to the inboard end of the console, reeve a sturdy string through it, and secure it to the hose by means of electrical tape or cable ties. This will restrict complete 360 degree rotation, thereby reducing the chance that the pressure gauge will unscrew, and will ensure that the console won't be lost during a dive, such as when entering the water or climbing up a ladder.

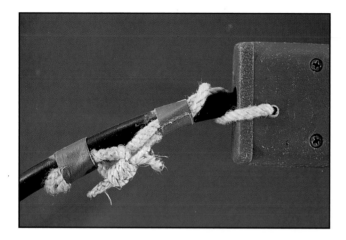

A Salt-free Diet

After every ocean dive, rinse your computer in fresh water, dry it, and store it in a water-free environment. Failure to rinse off sea water will allow salt to crystallize on the unit as the water evaporates, and will cause minerals commonly found in solution to precipitate on the metal contacts. Corrosion or encrustation makes it difficult to actuate the computer.

You'll also discover that a damp computer can be tricked into log mode or into scrolling no-deco times. It will beep annoyingly in your gear bag. Not only will the constant beeping drive you crazy, but it saps power.

Also, a computer bared to the spray, stored in a moist gear bag, or even exposed to the high humidity commonly found on boats, in fog, or around marinas, may not clear properly after a dive and may not display the surface interval.

When I was using Aladin Pros they sometimes refused to clear after a dive. Then, when I made the repetitive dive, they added the bottom time from the previous dive to the next dive. Internally, the computer was calculating correctly and even yielded an accurate decompression profile (although I didn't know it at first). It was only the bottom time *display* that was additive. I didn't rest easy until long after reaching the dock with no ill side effects. Beuchat offered no explanation for this bizarre behavior.

Computer Crashes and Contingencies for Safe Landings

Keep an eye on your computer even when it's not in use. If the low battery warning comes on, you want to know about it long before you're about to take the plunge. Which brings up an important issue: what to do if your computer crashes. (Dive computers always "go down.")

There are three times when a decompression computer can fail: before a dive, during a dive, and during surface interval. Obviously, if your computer breaks down or will not turn on prior to a dive, the default option is a set of decompression tables. That's why a computer diver must be well versed in working tables even though he doesn't expect to use them.

It is equally important to get in the habit of carrying back-up tables all the time. Top-of-the-line computers switch on automatically when they hit the water. Because of this, most people take it for granted that the computer is functional and don't bother to check it before hand. Not until they look at the display under water do they learn the horrible truth. It's a good idea to switch on your computer during your pre-dive check, just to be sure it's working.

A computer might malfunction during the conduct of a dive, possibly due to flooding or perhaps because you rapped your computer hard against a rock or wreckage. If this eventuality occurs, your situation might be complicated by the fact that you've already exceeded the no-decompression limit. Now you don't know your ceiling or ascent time.

When I first switched to computers I continued to carry two time pieces (a watch and an automatic bottom timer) and a set of decompression tables: an inexpensive back-up. I had occasion to use them, too. Now I carry two computers, an extravagance that most divers are not willing to bear. You have to decide for yourself how important it is to have a convenient and reliable way out of what could develop into an uncomfortable predicament.

If a computer shuts down after a dive, you lose the data needed to calculate a repetitive dive. The solution is simple: don't dive again until you have totally desaturated. Then buy or borrow another computer till yours is back on line. That's a bit like saying, "when you get sick, curl up in a corner until you get well." In technical diving, however, there are some dives that you just *have* to make. You might be on a week-long caving expedition to the Yucatan Peninsula, or on a multiday boat trip: premium investments in which every dive yields interest that might otherwise be lost.

If you *have* to dive again, take what information you know about your profile, estimate conservatively what you don't remember precisely, throw in a pound of chocolate fudge factors, and plug everything into a set of very conservative tables.

Envoi

Is this all there is to know about decompression tables, computers, and methodology? Of course not. The biggest hang-up in decompression theory is just that: it's all theoretical. Despite several centuries of observation and nearly a hundred years of actual study, our understanding of the physiological processes involved in inert gas absorption and elimination is still in the embryonic stage. Recent intense research has shown that much of what we think we know about the mechanisms of decompression and the incidence of DCI cannot be explained by actual case histories, or verified by statistical analyses.

Computers now on the market seek to compensate for some of these theoretical lapses and gaps in our knowledge with user actuated options. Workload can be factored into the decompression equation by taking air consumption into account (an adjunct of hoseless telemetry). Likewise, the program can be adjusted to consider the water temperature and actual ascent rate. With some models, you can even add arbitrary percentages of conservatism for age, weight, and so on.

Theoretically, it's possible to add so many additional fudge factors that your computer will never let you out of the water. The decompression program might describe an asymptotic curve that never eliminates the ceiling.

Then there's there apocryphal story about an astronomer in the 1960's who was asked by a science magazine to write a thousand-word article about life on Mars. He wrote five hundred times, "Nobody knows."

We've learned a lot more about Mars since then, and more about the physiology of decompression processes, but still no one has all the answers. We don't even know all the questions.

But we've got to go with what we have. If mankind had waited for the construction of the Concorde before he took to the air, the Wright Brothers would never have been allowed to fly. Human progress and the growth of thought are evolutionary developments. We aren't born with full-blown knowledge and understanding. We have to learn, and learn to improve.

Until someone invents a device that takes actual blood and tissue samples during all phases of a dive, or discovers a pill to prevent nitrogen absorption and bubble formation, the mathematical model that drives the modern decompression computer is the best we have.

I'll use it until something better comes along.

NITROX:
Beyond Air

One need not be a physicist to comprehend the underlying principles of "hyperbaric nitrogen-oxygen mixtures" and other "exotic gas blends." It will suffice to have some basic knowledge of chemistry and biology and, perhaps, the law of partial pressures.

In this and the next two chapters I'll explain what's necessary as I go along, and drop the appropriate buzz words as the concepts appear. First, some explanations.

The air we breathe consists of 20.9% oxygen, 78% nitrogen, 0.9% argon, and trace elements and molecules too minute to matter (carbon dioxide, hydrogen, helium, and carbon monoxide, among others). By definition, any other combination of elements in the third state of matter is a "mixed gas."

For simplicity's sake, the constituents in air are usually rounded off to 21% oxygen and 79% nitrogen (this figure includes argon which, like nitrogen, is an inert gas). The other 0.2% is inconsequential for calculation purposes.

Oxygen - the Good, the Bad, and the Ugly

The element in air that is necessary for sustaining life is oxygen. Oxygen in its natural state is diatomic. That is, two atoms of oxygen combine to form an oxygen molecule. Because of its affinity to share the electrons in its outer shell, free oxygen does not exist. The oxygen molecule is designated chemically as O_2 (pronounced oh-two). The subscript "2" indicates two atoms.

The human body can safely utilize oxygen only at specified partial pressures: too little partial pressure of oxygen (written ppO_2) causes hypoxia (you pass out), too much and oxygen becomes toxic. We can't live without oxygen, but sometimes we can't live with it, either.

Oxygen toxicity is a function not only of too high a partial pressure of oxygen, but of duration of exposure. This means that a high concentration of oxygen breathed safely for twenty minutes may induce detectable symptoms of oxygen poisoning after thirty or forty minutes.

According to accepted scientific parlance, partial pressures are given in atmospheres absolute - abbreviated ATA and pronounced as a two syllable word with each "A" being short. In pluralizing ATA (sometimes confusingly printed "ata"), if ATA's is more pleasing to your ear, let euphony guide your pronunciation, unless you're speaking in front of an audience of snooty hyperbaric physicians or nonmedical Ph.d's.

For durations up to an hour, the current accepted safe limit for the partial pressure of inspired oxygen lies between 1.4 and 1.6 ATA. This is equivalent to a depth of between 13 and 20 fsw *if you were breathing pure oxygen*. Breathing air, the partial pressure of oxygen reaches 1.6 ATA at a depth of 218 fsw.

How meaningful is this standard? I've been making 20-25 minute dives on air to 250 fsw for the past two decades without any ill effects, and have gone as deep as 290 fsw on air for fifteen minutes. Others have exceeded 300 fsw on bounce dives. A few have gone beyond 400 fsw on record breaking stunts (although some of them didn't come back).

Is this tempting fate? Is this cheating death? Or is 1.6 ATA a figure that is overly conservative? This is a controversial issue rife with outspoken dissidents who oppose all viewpoints that contradict their own. Standards vary. Military convention accepts 2.0 ATA for combat applications.

NOAA recommends spending no more than 45 minutes at 1.6 ATA. What is the true safe limit?

Let's go back to an example I've already used: what is a safe speed limit for vehicular travel on highways? To answer this question we first have to look at what criteria were used to select the national speed limit (aside from the gasoline crisis in the early 1970's, in which case the limit had nothing to do with speed but with the expedience of burning less fuel). However, let's *not* ignore limited access highways that are designed without sharp curves in order to facilitate high-speed travel.

The primary formula on which any safe highway speed limit is based is the lowest common denominator: that is, the highest speed at which it is safe for *everyone* to drive, including licensed senior citizens with cataract-riddled eyes and suffering undiagnosed senility. A healthy young athlete with lightning-like reflexes gains no favoritism for extraordinary ability. Neither is a Maserati given special treatment over a Model T Ford.

Navy divers are given an oxygen tolerance test in which they breathe pure oxygen in a chamber at a pressure equivalent to a depth of 60 fsw. In that case the partial pressure of oxygen is 4.8 ATA! Most pass the test without going into convulsions, without feeling twitches, without noticing any adverse effects.

Scientists claim that such a simulation is meaningless. For one thing the divers are at rest. The risk of toxicity is increased dramatically by a heavy work load. Furthermore, they contend that a person can make five hundred nonsymptomatic dives breathing gas with elevated partial pressures of oxygen, then suffer an oxygen hit on the five hundred and first. This is due to factors that are nowhere near to being understood.

If you are befuddled by this trenchant invalidation of standards, you have a right to be. There's no perfect answer. The ppO_2 limit you choose not to exceed is the one you feel comfortable with after acquainting yourself with the evidence, then by forming your own opinion.

Let me render assistance by adding some perspectives. Many people still believe that scuba of any kind is unsafe below 130 fsw. Recreational divers suffer DCI without exceeding the no-decompression limits. Some divers have gotten bent after decompression dives even though they followed conservative decompression tables and programs to the letter. There are cases on record in which technical divers have taken an oxygen hit while breathing oxygen in less than 20 fsw.

The best that can be said for now is that the majority of experts confer that partial pressures of oxygen somewhere between 1.4 and 1.6 ATA should be safe for everyone to breathe, with 1.4 ATA being favored because it offers a greater margin of safety. "Everyone" is intended to include the lowest common denominator: that is, the person with the least individual tolerance to oxygen toxicity.

Common consensus decrees that exposure to 1.4 ATA of oxygen for one hour should be safe for all. And *still* there is that conditional word "should." Drivers can lose control at any speed, and divers may still take hits. The best we can say is that 1.4 ATA of oxygen offers a greater latitude for safety.

"Caveat emptor," as the Latin saying goes. "Let the buyer beware."

There are two issues that everyone agrees on. One is that the adverse affect of high partial pressures of oxygen is cumulative and that therefore the ppO_2 should be reduced if longer durations are anticipated. The other is that the safe standards apply to conditions largely at rest, and that dives during which hard work is anticipated should factor in a lower ppO_2.

For the purposes of this chapter I am treating nitrox as a bottom mix in which the partial pressure of oxygen is less than 40%. I am also assuming adherence to no-decompression limits, in which case you can't stay down long enough to reach the point at which oxygen becomes toxic. Oxygen toxicity can lead to seizure and drowning.

For further discussion in this regard, read the sections on oxygen toxicity in the next chapter, "Accelerated Decompression." In the meantime, remember that rules and standards are created by people and are always subject to change. Nature, misunderstood though it is, honors laws that are immutable.

Nitrogen, Narcosis, and DCI

Like oxygen, nitrogen is diatomic. In chemical shorthand it is written N_2. The nitrogen in air is superfluous and serves no metabolic function. It is simply filler. At atmospheric pressure, nitrogen has no physiological effects, but at depth (that is, at increased partial pressure) it becomes narcotic. This narcosis is relieved immediately upon the reduction of partial pressure.

Decompression sickness was once called caisson disease because tunnelers who worked in submerged pressurized chambers called caissons were sometimes crippled upon return to the surface after long exposure to pressure. The hunched-over position of those afflicted led to the term "the bends." It is a complex disorder that ranges from mottled and itchy skin to mild tingling in the extremities to intense pain in the joints, and to a variety of enfeebling neurological conditions, some of which may be permanently disabling. It can also lead to death.

Somewhere in the evolution of the study of caisson disease it came to be called decompression sickness. This was later abbreviated to DCS. The words "disease" and "sickness" are misleading to the lay mind because both imply a malady of organic origin. Disease is something you catch, sickness is something you get.

Following the same pattern of wisdom, in recent years physicians, scientists, and hyperbaric specialists have arbitrarily changed the name of the ailment to decompression illness, and abbreviated it DCI. I have yet to find a doctor who can explain the difference between a sickness and an illness other than by a grin and a shrug of the shoulders. To me they seem the same.

With all due respect to the medical profession, I don't think of the bends as either a disease, a sickness, or an illness. It is not pathological in nature. The formation of nitrogen bubbles is a principle of physics, not of physiology. Nor is it an infection of bacteria or virus, or an abnormal cellular defect.

The bends is the result of an injury akin to a wound or physical trauma that is caused by an external event, such as a fall from a ladder or the blow of a blunt object. Call it an internal contusion if you will. That the injury occurred without breaking the skin makes no difference.

Therefore, I accept the common usage of DCI, but submit that it should stand as the abbreviation for decompression injury.

Nitrogen Tolerance

Susceptibility to oxygen toxicity and nitrogen narcosis varies according to individual tolerance,

in the same manner in which one person can drink alcoholic beverages all night with little noticeable effect while another is flat on his seat after two drinks. (I fall victim to the cheap date syndrome and stay away from the stuff.)

Likewise, there is some latitude in the susceptibility to DCI. (Read "Submersible Air Decompression Computers" for more in that regard.) But these are riddles for medical science to unravel. We just want to dive. (You can drive a car without understanding the principle of the internal combustion engine.) So let's put theory aside and get into practice.

What is Nitrox?

A combination of nitrogen and oxygen in any percentage other than that found in the atmosphere is called nitrox. This can refer to mixtures in which the percentage of oxygen is less than 21%, such as in undersea habitats and saturation diving chambers, but usually describes a blend in which the percentage of oxygen is higher. In the latter case, "enriched air" is the more descriptive term, and "enriched air nitrox" (EANx) is the formal term, while "nitrox" is used commonly and "enriched air" is understood.

The commercially blended nitrox mix in most general use consists of 32% oxygen, the balance nitrogen, and is officially designated NOAA Nitrox I, or NNI. Generically this blend is written nitrox-32, formally it is written EAN-32. (NOAA Nitrox II consists of 36% oxygen, the balance nitrogen, and can be written nitrox-36 or EAN-32.) Nitrox can also be custom blended to yield any percentage of oxygen that is desired.

Nitrox is generally manufactured by adding air to oxygen in proportions that yield the desired percentages. This means that nitrox contains the same impurities and trace gases that are contained in air. "Pure" nitrox can be made by mixing oxygen with nitrogen, each in distilled form, but no advantage is obtained for breathing purposes because the percentage of impurities and trace gases in air is so small and because their physiological effect is negligible.

Nitrox tastes no different from air, smells no different, and breathes no different. A cylinder full of nitrox is therefore difficult to distinguish from a tank of air. The gas must be analyzed with an oxygen sensor in order to distinguish one from another.

Label your Gas

Because nitrox cannot be differentiated from air without oxygen analyzing equipment, certain precautions must be taken in handling. The dedicated tank is the safest and most obvious way to store and carry oxygen enriched air. In its purest form the cylinder is painted yellow with a green shoulder, and has the word "NITROX" stenciled in capitals along or around the body of the tank.

The loudest proponents of the dedicated tank system are - you guessed it - those who sell tanks and offer nitrox fills. Some shops are so adamant about their position that they refuse to fill tanks that are not dedicated. To be sure, bright yellow and green tanks are highly visible and not likely to be mistaken for a tank of air.

The problem with dedicated cylinders is that they should remain dedicated. That is, they should no longer be used for air because of the possibility of confusion. As will soon become apparent, breathing air when you think you're breathing nitrox can have deleterious side effects. The same is true if you take a tank of nitrox too deep because you think it's filled with air.

As an alternative to the dedicated cylinder, a prominent bright green and yellow nitrox label can be stuck onto an ordinary scuba tank close below the shoulder. The tank is henceforth not only designated, but dedicated, except that the label can be removed.

At a minimum a dive shop gas fill sticker will suffice. A fill sticker is much like a VIP sticker except that instead of having a hole punched through a date, blank spaces are provided so the filling station operator can inscribe with an indelible marker the actual blend with which the tank was filled. The fill sticker can be universal in that it can also be used for other kinds of gas fills, such as helium.

Oxygen Cleaned Tanks - Fact or Fallacy?

Another bone of contention among nitrox users is whether nitrox tanks need to be cleaned for oxygen service. Some say yes, some say no. Depending upon the circumstances both camps can be right - but the nays are right more often than the yeas. The difference is not in how nitrox tanks are used but in how they are filled. Follow my reasoning.

Although oxygen doesn't burn and isn't explosive, it supports combustion. By definition, fire (or burning) is a persistent chemical reaction between a combustible substance and oxygen; or, the rapid oxygenation of a material. By contrast, rust is the oxygenation of iron that is being converted to ferrous oxide - a slow burn, if you will.

Cylinders intended for the storage of oxygen must be cleaned of all contaminants and oily residues (hydrocarbons). This is because those contaminants and oily residues take on an explosive quality in a pure oxygen environment. If these substances reach the flash point or encounter a source of ignition, a fire will erupt. The fire may burn slowly or it may lead to an explosion. The catch phrase here is "pure oxygen environment."

According to the Compressed Gas Association (CGA), any cylinder or device that sees service in an atmosphere in which the oxygen content is greater than 40%, must be oxygen cleaned. This means that blends as concentrated as nitrox-40 can be carried in tanks without special cleaning. These blends include nitrox-32, nitrox-36, nitrox-40, and other custom blends whose percentage of oxygen is 40% or less. If you accept a strict line of demarcation, only those tanks used to carry nitrox-41 and higher must be cleaned for oxygen service.

These CGA standards apply to valves and regulators, too.

As I wrote in the previous section, those who cry the loudest over the need for oxygen cleaned equipment are those who have a vested interest in supplying those items and services for profit. Notwithstanding this observation, sometimes they cry with a voice of reason.

If you have your tanks filled with nitrox from a storage bank, in which the gas has been pre-mixed, you have nothing to worry about as long as the blend is nitrox-40 or less. This is the usual manner in which technical shops fill their customers' tanks. But if your cylinder is used as the vessel in which to mix oxygen with air in order to create the desired blend, then the tank and valve should be oxygen cleaned no matter what percentage of oxygen is contained in the final blend.

Oxygen becomes potentially explosive at pressures higher than about 2,400 psi (the industry standard established by the CGA). For that reason, when oxygen is mixed with air to create high-pressure nitrox, the oxygen is pumped in first and is then topped off with air to reach the desired final pressure. In such a case, during the time that oxygen is introduced into the empty tank, the partial pressure of oxygen is 100%.

You can't put air in the tank and then top off with oxygen because that would necessitate pumping oxygen at a pressure that is higher than industry standards.

Oxygen Cleaning Tanks and Valves

Lest you remain in a quandary about the arcane process of cleaning equipment for oxygen service, here's how it's done. Use clean tools only - not auto mechanic tools that are covered with grease. Do *not* use trichloroethylene as a cleansing fluid. This freon-based chemical was removed from the market because it was found to be environmentally unfriendly: it contributed to destroying the ozone layer (it might also be carcinogenic).

Depressurize the tank and unscrew the valve. Mix a solution of DOT-113 with a concentration of 5-10%. (DOT-113 is sold as a powder; it dissolves better in hot water than in cold. Too high a concentration produces too many suds which are hard to rinse away.) Pour the liquefied solution into the tank until it is two-thirds full, then screw in a plug. Lay the tank on its side and rotate it (that is, roll it around its length) for about twenty minutes in each direction. A motorized tank tumbling unit does this the best.

Dump the liquid down the drain or use it in another tank that day (but don't save it overnight). Using a toothbrush, scrub the internal neck threads with cleaning fluid. (This is because the plug prevents all the threads from being cleaned.) Flush the tank thoroughly with *hot* tap water until all traces of the chemical and suds are washed out. Stand the tank upside down and let drain.

Blow hot air up through the neck into the tank and force-dry it. This is critically important for steel tanks in order to prevent rusting. The best way to generate hot air is from a heat gun or industrial air dryer. A large wattage hair dryer

will suffice. In a pinch you can use the exhaust or hot air discharge of a household vacuum cleaner, but you must make sure that the air is clean - no dust. If you work in your garage, prevent roaches and spiders from crawling inside the tank and laying eggs.

The valve assembly must also be cleaned for oxygen service. Disassemble the valve and remove all non-metallic items such as the o-ring and inner seat. Soak the block, spring, and other metal parts in ordinary white vinegar. Half a day or less should remove all signs of corrosion. Rinse the parts with tap water, let them dry, and soak them in the same solvent you used to clean the tank interior. Scrub all surfaces with a toothbrush and use a cotton swab to clean out the orifice. Use the same rinsing procedure as that used for tanks. Dry in air.

Replace the original seats and o-rings with those made of viton. Different types of o-rings may be indistinguishable once removed from the original packaging, so be careful not to get them confused.

Lubricate all parts with a medical oxygen lubricator such as Christo-Lube or Krytox grease. Do *not* use petroleum-based lubricants such as silicone grease. Reassemble the valve and screw it into the tank. Now it's safe to fill the tank with oxygen or nitrox.

Paint or label the tank so you don't accidentally recontaminate it. Fill the tank with air (as opposed to nitrox) only if the air is pumped through an oil-free compressor or if there's enough filtration to guarantee clean-grade air that won't compromise the oxygen-cleaned integrity of the tank.

Depth Limitations of Nitrox

Now that we've covered the mechanics of nitrox, let's get into usage.

Nitrox is strictly a shallow water gas. As noted above, because nitrox contains a higher fraction of oxygen than air, the possibility of incurring oxygen toxicity is encountered at much shallower depths. For nitrox-32, 1.6 ATA of oxygen is attained at 130 fsw. This is the recommended depth limit for a no-decompression dive for that particular concentration.

Taken with the admonitions discussed in previous sections of this chapter, remember that this border line is a natural physiological (and somewhat arbitrary) boundary, not an artificial restraint adopted by overly restrictive, anti-progressive training agencies.

For nitrox-36, the maximum recommended depth is 110 fsw. Nitrox blends in which the percentage of oxygen is more than 21% but less than 32% can safely be breathed at depths greater than 130 fsw, but the advantage is not as great.

The precise depth limitation for any nitrox blend can be calculated with this equation:

$$P = ppO_2/fO_2$$

P refers to pressure (atmospheres absolute). The variables in the equation are the partial pressure of oxygen (ppO_2) in atmospheres absolute and the fraction of oxygen in the mix (fO_2). The ppO_2 is arbitrarily assigned. If we use 1.6 ATA then divide it by .32 (the decimal equivalent of oxygen) we get the whole number 5 (which is atmospheres absolute).

Convert P to feet of sea water with this conversion formula:

$$fsw = (P - 1) \times 33$$

From 5 subtract 1 atmosphere absolute (for surface atmosphere) then multiply by 33 (each atmosphere equals 33 fsw) and the answer is 132 fsw.

Less Nitrogen, not More Oxygen

The advantage of breathing nitrox comes not from the increased percentage of oxygen, which we have seen has it own potentially harmful side effects, but from the decreased percentage of nitrogen. Less nitrogen equates to less inert gas absorption, which translates to extended no-decompression limits and to shorter decompression times than those required for equivalent depths and durations breathing air.

The whole point of breathing nitrox is that you can stay longer and hang less.

The physics and physiology of breathing nitrox are the same as they are for breathing air. No extra training is required in order to master the principles of nitrox diving, only an understanding of the underlying differences between nitrox and air.

Equivalent Air Depth

There are two popular ways to figure no-decompression limits and decompression times. Both are related to dive tables. For those of you who are mathematically inclined, and like to fool around with figures, try this formula:

$$EAD = \frac{(1.0 - fO_2)(D + 33)}{.79} - 33$$

The concept of "equivalent air depth" (EAD) equates the partial pressure of nitrogen in nitrox to the partial pressure of nitrogen in air, and lets you use tables that you're already familiar with for determining no-decompression limits and for calculating decompression. The two variables in the formula are fO_2 (the fraction of oxygen in the mix) and D (the depth in feet of sea water).

Note that .79 is the *total* inert gas content of air, including nitrogen and trace gases. Also note that this same formula can be used for *any* combination of nitrogen and oxygen. Because nitrox-32 is the blend that is most widely breathed, and is commercially available, I'll use it to demonstrate a sample problem.

In the formula above, (1.0 - fO_2) becomes (1.0-.32) which equals (.68). Choose a depth and we're ready to go. Let's say D is 80 fsw. (D + 33) becomes (80 + 33) which equals (113). By multiplying .68 by 113 we get 76.84. Dividing that by .79 yields 97.27. Forget the decimals and subtract 33, and the result is 64. Because tables are laid out in ten-foot increments, we round up to the next deeper depth in tens, and get an equivalent air depth of 70 fsw.

This means that for decompression purposes a dive to 80 fsw on nitrox-32 is equivalent to a dive to 70 fsw on air. According to Navy Tables, the no-decompression limit is increased from 40 minutes to 50 minutes - you get an extra 10 minutes bottom time for free!

By the same token, a 130-foot dive on nitrox-32 is equivalent to a 110-foot dive on air - you still get ten more minutes for free. A 100-foot dive on nitrox-36, in which the fraction of nitrogen is .64, comes out to 74.74 fsw, which rounds up to 80 fsw.

A more compelling example can be shown by breathing nitrox-39 at 100 fsw, in which the equivalent air depth is 70 fsw - a 30-fsw differential. You can optimize bottom time by creatively adjusting the fraction of oxygen in the blend.

Here's another equation that yields the same answer:

$$EAD = \frac{[fN_2 \times (fswA)]}{.79} - 33$$

Although this equation looks dissimilar to the previous one, it's verily alike. The designations have been changed to protect the innocent.

The fN_2 is the fraction of nitrogen in the mix. If we continue to use nitrox-32 as our example gas, you will note that fN_2 is the same as (1.0 - fO_2), or .68. Also apparently different is fswA, which stands for "feet of sea water absolute." This is just another way of writing (D + 33 fsw). Use the formula that suits your taste.

Repetitive EAD

Standard residual nitrogen groups are used for calculating repetitive dives, the same as they are for air. In the first example, you would take a group designation letter from the surface interval durations in the row for a 70-fsw dive, instead of for an 80-fsw dive, and use that to calculate your repetitive dive. What could be simpler?

NOAA Nitrox Tables

Alternatively, a special set of tables has been designed by NOAA for nitrox-32 that is identical in format and use to the U.S. Navy Standard Air Decompression Tables. Since you've already learned to use dive tables in your basic scuba course, you'll have no difficulty applying nitrox tables. The extended bottom times are built in.

For example, the no-decompression limit breathing air at 130 fsw is 10 minutes; for nitrox-32 it is 20 minutes. Thus the decrease in nitrogen more than doubles your bottom time. I say "more" because if it takes two minutes to reach the bottom, your effective working time - or playing time - is 18 minutes instead of 8.

Repetitive dive nitrox tables also follow the same format as Navy Tables, and offer more bottom time than equivalent air dives.

NOAA Tables or EAD?

EAD has some advantages over NOAA nitrox tables. Most people already own a set of dive tables: there's nothing to buy, there's no chance of mixing up your tables and using the wrong one, and there's nothing extra to search for while preparing for a dive. You only need to know the equivalent air depth for the blend you are using, and you're ready to calculate.

If you rely on nitrox tables and you're the only one in the group using nitrox, there may not be a set to borrow if you forget to bring your NOAA tables.

On the other hand, if you do use nitrox tables, make very sure that another diver doesn't pick

NOAA/Navy EAD

(all depths in feet of sea water)

Actual Depth	Nitrox-32 EAD	Navy Air Table (Rounded Up)
40	29.84	30
50	38.44	40
60	47.05	50
70	55.66	60
80	64.27	70
90	72.87	80
100	81.48	90
110	90.09	100
120	98.70	100
130	107.30	110

them up accidentally in order to calculate an air dive. The consequences could prove serious.

What is the flaw in the concept of NOAA tables? If you've read the chapter on decompression computers, you've already guessed it. The safety component is marred by the tables on which the decompression model is based. I've already gone to great lengths to demonstrate the fallibility of Navy Tables with respect to technical diving. Since NOAA nitrox tables are a direct derivative of Navy Tables, they suffer from the same inherent lack of conservatism.

Table Idiosyncrasies

Now let me show you some quirks in the Navy and NOAA tables, and the reasons I do not endorse them.

Oddly enough, when breathing nitrox, a decompression dive has a wider margin of safety than a no-decompression dive. The reason is that on decompression dives you hang according to air tables that assume you're breathing air. The lower ppN_2 in nitrox is in your favor since your nitrogen absorption during decompression is less. With less nitrogen being inhaled you can off-gas faster. This applies not only to decompression dives using the EAD method, but also when using NOAA nitrox tables because they are based on the EAD concept.

The longer the decompression,
the wider the margin of safety.

On the down side, remember that Navy Tables totally desaturate you after only twelve hours - far too short in my mind. The same is true for NOAA nitrox tables.

Refer to the NOAA/Navy EAD Table while I plug some numbers into the EAD formula and demonstrate the inconsistencies in the no-decompression limits.

Notice that at the 110-120 fsw break point both depths yield an EAD of 100 fsw. Obviously, one has to be less conservative than the other. Because the 110 fsw EAD works out to 90.09 fsw, and because according to Navy Table rules the depth must always be rounded up to the next deeper depth, you get an incredible margin of safety - practically 10 fsw.

However, the EAD for 120 fsw is 98.70, which yields a safety margin of only 1.3 fsw. (Even that, of course, is better than "pure" Navy Tables when breathing air.)

Watch the pattern that evolves. The 40 fsw EAD can be discounted because it equates to 30 fsw on air, where the no-decompression time is unlimited. The 50 fsw EAD has a close margin that gradually widens as it approaches the 110 fsw EAD, from which point the EAD flips over the next ten-foot increment to the 120 fsw EAD where it again becomes close.

This means that when using Navy and NOAA nitrox-32 tables, the most conservative depth is 110 fsw - at least, according to fudge factor guidelines that I set forth in the previous chap-

ALLOWABLE EXPOSURES TO OXYGEN

Partial Pressure of Oxygen in Atmospheres Absolute	Maximum Continuous Exposure		Maximum Daily Exposure	
	Minutes	Hours	Minutes	Hours
1.6 ATA	45	.75	150	2.5
1.5 ATA	120	2.0	180	3.0
1.4 ATA	150	2.5	180	3.0
1.3 ATA	180	3.0	210	3.5
1.2 ATA	210	3.5	240	4.0
1.1 ATA	240	4.0	270	4.5
1.0 ATA	300	5.0	300	5.0
0.9 ATA	360	6.0	360	6.0
0.8 ATA	450	7.5	450	7.5
0.7 ATA	570	9.5	570	9.5
0.6 ATA	720	12.0	720	12.0

ter. By extension, decompression dives made at 110 fsw are equally as conservative. The least conservative depths are 50 fsw and 120 fsw.

Of course, by using these tables you still have the additional safety factor of a square profile when you may actually dive in a multilevel fashion. When in doubt, make a five or ten minute safety stop at 10 fsw. Since you'll be breathing nitrox you'll be doing yourself an EAD favor.

Let me emphasize, however, that overall a nitrox-32 dive to 110 fsw is no more or less conservative than a nitrox-32 dive to any other depth or duration. The empirical decompression profile is a curve. It is the way Navy and NOAA tables are printed - squared off in increments of five, instead of in true and accurate values - that create the disparities noted above.

Alternative Tables

After spending so much space debunking Navy and NOAA tables, the reader might wonder why I used them to exemplify the highlights and pitfalls of nitrox diving. The reason, as I stated in the previous chapter, is that they are universally available. In that regard they make excellent reference materials. But this doesn't mean that I promote their use in the water.

Better, more conservative tables exist. Once you master the EAD concept it can be applied to *any* set of air tables. Bring on the British and Canadian algorithms, and the print-outs of Buhlmann and Huggins. Some certifying agencies have developed their own proprietary Doppler tables: so called because the decompression profiles were tested empirically by means of an ultrasonic Doppler, an instrument which detects bubbles flowing through the circulatory system. All these tables are recommended.

The Royal Navy uses three standardized nitrox mixes (32.5%, 40%, and 60% oxygen) and has tables for each. Better yet, according to Richard Vann, the French decompression tables "provide EAD lists for mixtures ranging from 25% to 50% oxygen in 5% increments. . . . These mixes can be used with any decompression table." (*Workshop on Enriched Air Nitrox Diving*, 1989.)

Another advantage of using the EAD concept and alternative tables instead of NOAA tables is that you can switch back and forth between nitrox and air on repetitive dives, or between various nitrox blends. One table serves all.

Nitrox Computers

With the popularization of nitrox, it didn't take long for computer manufacturers to redesign their programs in order to develop a wet model that could calculate decompression profiles attuned to different percentages of oxygen. The algorithms themselves didn't need modification, only the function in the equation that defined the relationship between gases.

Already you must be thinking that you have to buy another computer for every percentage of oxygen you'd ever want to breathe. Not to fear. Farsighted industrialists vaulted a quantum leap beyond such a simple expedient. They designed a nitrox computer that can be programmed by the user in the field.

This model has all the extras of a full-function air computer, in addition to which it allows you to specify the percentage of oxygen on which you want the decompression profile based. Your selection can be made at a moment's notice, and chosen incrementally to reflect any blend it is possible to mix.

You can't make changes under water. You have to pre-set the computer on the surface so it concurs with the blend that you have in your tanks. Then follow the profile that is shown on the display, the same as you would on an air dive with an air computer.

If you haven't bought a computer yet or are looking for a second one in order to have a backup, this is the one to buy. Not only is it adaptable to every blend of nitrox, but by programming the oxygen content for 21% the unit calculates for air.

Formula for Success

By now you must have already guessed that anyone who can dive on air can learn to dive on nitrox. The concept is simple to grasp, the principles easy to master.

I hate to complicate the issue now, especially with dispensable data, but those who have an inquisitive mind might like to fool with equations. Skip this section if you just want to buy nitrox and dive without complications.

All three equations that follow are related. They have simply been rearranged in order to solve for a different variable.

$$fO_2 = \frac{ppO_2}{P} \qquad ppO_2 = fO_2 \times P \qquad P = \frac{ppO_2}{fO_2}$$

P is the pressure absolute, or atmospheres absolute (ATA). It is the same as (fswA) in the equation given several sections above. To get P you must divide (D + 33) by 33, in which D = depth. Read fO_2 as the fraction of oxygen. And ppO_2 is the partial pressure of oxygen.

If you want to make custom blends you can figure the best mix by solving for fO_2. For example, suppose you're working a site that's 70 fsw deep. First solve for P by adding 70 and 33, then divide 103 by 33. The answer is 3.12. If you want your partial pressure of oxygen to be 1.6 ATA, divide that number by 3.12. Nitrox-51 is the optimal blend to breathe.

(Remember that for this high a concentration of oxygen the tanks should be cleaned for oxygen service. Remember too that a partial pressure of oxygen as high as 1.6 ATA should not be breathed on dives during which hard work is anticipated.)

Now go back to the EAD formula (either one). The equivalent air depth is 33.49 fsw. If you're using a table, round up to 40 fsw. If you're using a nitrox computer, program it for the precise blend. Either way, your no-decompression time will be increased enormously.

Once your tanks are filled with that particular blend, those tanks can't be used on dives to deeper depths, only shallower. If the 70 fsw dive fails to materialize and deeper dives come your way, you either have to dump the gas or have additional tanks available so you can store the nitrox-51 for another diving opportunity.

Keep in mind, too, that the maximum duration permitted for breathing the extreme allowable partial pressure of oxygen - by the least conservative table - is only forty-five minutes. You'd be better off lowering the ppO_2 to 1.5 ATA or 1.4 ATA.

The catchphrase in technical diving is "Watch your ppO_2's."

If you work out the math for 1.5 ATA, and breathe nitrox-48, the EAD is 29.58 fsw. Work it out for 1.4 ATA, breathing nitrox-45, and the EAD is 25.67 fsw. In both instances, by reducing the fraction of oxygen you significantly increase your safety margins, because at those equivalent air depths your permissible oxygen dose (for this example) becomes, for all practical purposes, infinite.

Let's look at it another way. Instead of breathing nitrox-32 at the limit of allowable tolerance (130 fsw), let's create a blend based on 1.5 ATA. The resultant percentage of oxygen is 30% (fraction of oxygen equals .30). The actual EAD is 111 fsw. If you figure your decompression on a table, you get caught in the round up just like a herd of cattle. The effective EAD becomes 120 fsw.

This shortens your no-decompression time, but gives you a greater margin of safety with respect to oxygen toxicity, thus reducing the risk of seizure and unconsciousness. The last thing you want to do under water is to stop passing gas.

The purpose of this discussion is to illustrate the point that "best mixes" for depth are not necessarily the best mixes for duration. If you're going to mess around with custom blends, don't push the envelope. Use a partial pressure of oxygen that provides a generous margin of safety.

The Proof is in the Pudding

The first time I used nitrox scared my booties off. The tag on the tank that said nitrox-32 did nothing to dispel years of familiarity with accepted decompression penalties. After a long dive and a short hang, the EAD tables told me to come up. But my trusty computer - admittedly more conservative than Navy Tables, and making calculations based on air as a breathing medium - said I still had twenty-three minutes to decompress.

Intellectually I understood the principle of reduced nitrogen content, and I accepted the validity of the concept of reduced decompression, but my gut feeling was that it was all a pile of turkey turds. I exercised great power of will and forced myself to surface. Then I sat on the boat fully dressed, ready to jump back in the water at the first tingle or sign of pain or numbness.

I suffered no ill effects other than excessive sweating, not all of which was brought about by the hot sun and sultry air. It worked for me; it will work for you.

Nitrox Availability and Certification

You can buy nitrox of any desired blend from commercial gas suppliers, but they sell it only in large storage cylinders. Furthermore, that leaves you with the problem of transferring the gas to your scuba tank. For that task you need special hose connectors and either a cascade system or a booster pump.

EAD in fsw=	30	40	50	60	70	80	90	100	110	120	130	140
Nitrox-28	36	47	58	69	80	90	101	112	123	134	145	155
Nitrox-29	37	48	59	70	81	92	103	114	126	137	148	149
Nitrox-30	38	49	60	71	83	94	105	117	128	139	143	
Nitrox-31	39	50	62	73	84	96	107	119	130	137		
Nitrox-32	40	51	63	75	86	98	109	121	132			
Nitrox-33	41	53	64	76	88	100	112	123	127			
Nitrox-34	42	54	66	78	90	102	114	122				
Nitrox-35	43	55	67	80	92	104	116	117				
Nitrox-36	44	57	69	81	94	106	113					
Nitrox-37	46	58	71	83	96	108	109					
Nitrox-38	47	60	72	85	98	105						
Nitrox-39	48	61	74	87	100	102						
Nitrox-40	49	63	76	89	99							

EAD CONVERSIONS

In the left column are listed the most common nitrox blends breathed on the bottom in place of air. Across the top row are shown equivalent air depths in feet of sea water. The values shown in the body of the table are actual depths in feet of sea water. To determine equivalent air depths for use with any decompression table, match the chosen nitrox blend with the planned actual depth or the next deeper depth, then use the value shown at the top of that column.

For example, for a dive to 120 fsw breathing nitrox-32 there is no exact actual depth shown. The next deeper depth is 121 fsw. The EAD shown at the top of that column is 100 fsw.

Alternatively, to determine the optimum nitrox blend for a particular actual depth, correlate the planned depth shown in the body of the table with the various nitrox blends listed in the left column. By this means, for example, you will note that a dive to 100 fsw breathing nitrox-39 yields an equivalent air depth of 70 fsw - an EAD advantage of 30 fsw. Utilize this method for determining best custom blends.

In producing this table from the EAD formula, calculated depths were rounded up to the next whole number. Nitrox blends in which the percentage of oxygen is less than 28% lie in the realm of diminishing returns, and are seldom used.

Note that the values shown to the right of the thick jagged border represent depth limitations based upon a partial pressure of oxygen of 1.6 atmospheres absolute. These values are less than values based upon EAD calculations. For example, the EAD for nitrox-32 is 133 fsw, but the ppO2 limitation shown here is 132 fsw. This accounts for the apparent disparity in the sequential increase of values.

Nitrox-	28	29	30	31	32	33	34	35	36	37	38	39	40
1.6 (45 min)	155	149	143	137	132	127	122	117	113	109	105	102	99
1.5 (120 min)	143	137	132	126	121	117	112	108	104	100	97	93	90
1.4 (150 min)	132	126	121	116	111	107	102	99	95	91	88	85	82
1.3 (180 min)	120	114	110	105	101	97	93	89	86	82	79	77	74
1.2 (210 min)	108	103	99	94	90	87	83	80	77	74	71	68	66
1.1 (240 min)	96	92	88	84	80	77	73	70	67	65	62	60	57
1.0 (300 min)	84	80	77	73	70	67	64	61	58	56	53	51	49

ppO2 DEPTH LIMITS

This table enables the diver to determine any one of three variables at a glance, instead of having to solve the equation for each variable. The left column lists the partial pressures of oxygen in atmospheres absolute, with the total recommended single exposure given in parentheses. The top row shows the fraction of oxygen in the most common nitrox blends. The values shown in the body of the table are actual depths.

This table was created with calculations made from the following formula: $P = ppO_2/fO_2$. For example, a ppO2 of 1.6 ATA divided by .4 (the fraction of oxygen in nitrox-40) yields 4 ATA. To convert P (pressure absolute) subtract 1 ATA (surface atmosphere) and the result is 3 ATA, or 99 fsw. (1 ATA equals 33 fsw.)

1) O2 time limit ($ppO_2 = fO_2 \times P$) - To find the O2 time limit for a specific nitrox mix at a specific depth, locate the fraction of oxygen (fO2) for the mix across the top row. Go down that column to the exact depth or next greater depth. Go left across this row to the left column to find the partial pressure of oxygen (ppO2) and oxygen time limit for a single dive.

2) Best mix relative to oxygen component ($fO_2 = ppO_2/P$) - To find the best mix for a specific depth, locate the oxygen time limit and corresponding partial pressure of oxygen (ppO2) desired for the dive. Go right across the row to the planned depth (P) or the next greater depth. Then go to the top of that column to find the best mix (fO2).

3) Maximum depth ($P = ppO_2/fO_2$) - To find the maximum depth for a specific nitrox mix, first locate the oxygen time limit and corresponding partial pressure of oxygen (ppO2) in the column at left. Next locate the fraction of oxygen (fO2) of the nitrox mix across the top row. The intersection of the partial pressure of oxygen (ppO2) row and the fraction of oxygen (fO2) column gives the maximum depth (P) for the dive.

Usage note for BEST MIX

To choose the best nitrox mix for a specific depth, find that depth in the body of the EAD conversion table (or the next deeper depth if the precise depth is not listed). On that table, the farther a column stands to the left, the shallower the EAD. For a specific EAD there will be several nitrox mix choices. Transfer those choices to the ppO2 table and solve for the partial pressure of oxygen. If more than one ppO2 is offered, take the lower ppO2.

For example, you want to dive to 110 fsw. The EAD column for 90 fsw shows four mixes that can be breathed at that depth: nitrox-33 through nitrox-36. The ppO2 table shows that nitrox-36 and nitrox-35 have a ppO2 of 1.6 ATA and a single oxygen exposure limit of 45 minutes, whereas nitrox-34 and nitrox-33 have a ppO2 of 1.5 ATA and a single oxygen exposure limit of 120 minutes.

The best mixes for 110 fsw are nitrox-33 and nitrox-34, because they offer more bottom time with a lower ppO2.

Alternatively, you can blend your own nitrox by first putting oxygen into your tank and then topping off with air at your local dive shop.

Both of these methods are meant for advanced technical divers only, and will be discussed more fully in the chapter on "Expedition Diving." These procedures require a greater fund of knowledge, a better understanding of decompression theory, and more expertise than the regular nitrox diver needs to have.

For everyday purposes and common blends, a growing number of dive shops now offer nitrox fills to certified nitrox divers. The typical certification course consists of three hours of lecture - which reiterates the information that is given in this chapter - and an official certification dive. No big deal.

Once you've been issued a card you can either rent dedicated tanks from a participating dive shop or dedicate your own. Then just slap on your regulator - it needs no special cleaning or adapting - and breathe long and deep.

Dedication is Forever

It's extremely important to remember that if you fill non-dedicated tanks such as your own scuba cylinders with nitrox, they must be completely drained before you can use those tanks to make a dive on ordinary air. Otherwise, by topping off with air, the residual high concentration of oxygen will upset the gas balance of the subsequent fill. If you exceed the safe depth of the resultant blend that's in the tank, it could bring you into the realm of oxygen toxicity.

It's also important to maintain the decontamination of nitrox tanks. Once cleaned for oxygen service, don't have those tanks filled from a station whose gas isn't certified contamination free. If the shop doesn't use an oil-free compressor or if it doesn't pump air through adequate filtration to block contaminants and oily residues, you could introduce enough impurities into the cylinder to compromise its integrity. Then the tank must again be oxygen cleaned or you risk the danger of fire and explosion.

Cheap at any Price

Nitrox fills cost more than air fills. This is due partially to the purchase price of oxygen, but is attributable also to the capital investment entailed by the proprietor in establishing a nitrox blending, storage, and cascade system complete with attendant hose connectors, and the labor involved in mixing and dispensing the gas.

You get what you pay for. When you pay for nitrox, you get more time on the bottom. This increase in bottom time - the most important aspect of diving - is well worth the additional cost of nitrox. In fact, that cost is negligible when compared to all the other expenses that a dive necessarily incurs: gasoline, highway tolls, entrance fees, boat charters, air fare, lodging, food, and so on.

Fringe Benefits

For extending bottom time and for shortening decompression on shallow dives, nitrox is the way to go. There are other bonuses, too.

One beneficial adjunct that results from breathing a nitrox blend in which the percentage of nitrogen is reduced is a consequent reduction in narcosis. Folks who have an unusually low tolerance to the narcotic effects of nitrogen will detect a difference that is more than superficial. (Decreased narcosis is a concept we will discuss in greater detail in the chapter on "Mixed Gas.")

Nitrox breathers have also reported that they have more energy and suffer less post-dive fatigue after a nitrox dive than after an equivalent air dive. This may be due to the elevated partial pressure of oxygen rather than to the reduced partial pressure of nitrogen, but who cares? As long as it works.

Nitrox can be utilized on deeper dives too, as a travel gas and as a way of accelerating decompression. Let's turn to another type of technical diving in which nitrox plays a large and important role.

ACCELERATED DECOMPRESSION
Breathing Nitrox and Oxygen

Once after a lengthy decompression someone asked out of curiosity how long I had hung. As I frowned, and fumbled through my gear box for my timepiece in order to answer the question, my good friend Trueman Seamans said with feigned sarcasm, "You don't need a watch, you need a calendar."

Alas, many times since I have wished that his statement contained less truth. But the fact of the matter is that if you want to buy more time on the bottom you have to pay for it with decompression. Protracted, often cold and strenuous decompression is the baneful reality of technical diving.

Worse yet, at depths in the range of 200 fsw and beyond, the exponential curves on which decompression tables and computer algorithms are based rise sharply in amplitude.

For example, according to Navy Tables, the least conservative decompression model of them all, a 250-fsw dive for twenty minutes requires an hour of decompression. Spend an extra five minutes on the bottom and the decompression penalty becomes more than an hour and a half: thirty-two minutes additional hang time for five more minutes of fun.

Transfer this scenario to decompression computers, with their greater conservatism, and you pay the price with interest and with double digit inflation. The feasibility of long, deep dives begins to wane.

The Bad Gas

Everyone knows that the culprit in air that's responsible for decompression is nitrogen. (Some scientists think that oxygen may also play a part in it.) Once this was established, it didn't take long to reason that by reducing the percentage of nitrogen in the breathing medium, the amount of nitrogen that went into solution in the blood and tissues was also reduced. Thus the rationale for nitrox blends, described in the previous chapter.

But the seesaw effect of reducing nitrogen increases the oxygen content. This change in balance can aggravate the risk of incurring oxygen toxicity, thus limiting the depth to which the mixture can be breathed.

Replacing some or all of the nitrogen with other inert gases, the subject of the next chapter, merely replaces nitrogen absorption with alternative inert gas absorption, with no effective gain as far as decompression is concerned. But what happens if you breathe air during the working phase of the dive, then switch to another gas during the decompression phase - a deco gas with less, or no, nitrogen?

I won't lead the horse by the cart: oxygen decompression has been around for a long time.

Oxygen in the Military

Both the Navy and NOAA diving manuals have a table for what they call "surface decompression using oxygen" (Sur-D O_2 in the biz). Before you conjure up images of a diver relaxing in a bunk on a surface support vessel and decompressing by inhaling oxygen through a hose and trickle valve, let me explain that "surface decompression" is done in a deck decompression chamber.

The procedure requires an ascent rate of 25 fsw per minute until the diver reaches his appropriate ceiling, followed by partial in-water decompression on air. When he leaves the 30-fsw stop, however, instead of going to the 20-fsw stop he comes straight to the surface.

His tenders rip off his tanks, regulators, and weight belt (or helmet and lead shoes, as the case may be) and shove him into the chamber, which is then pressurized to 40 fsw. The elapsed time from 30 fsw in the water to 40 fsw in the chamber should not exceed five minutes, the interval allowed before the probable onset of DCI.

The diver then breathes pure oxygen through a mask, and can serve his decompression penalty in comfort. Sounds appealing.

In-water Oxygen Decompression

Technical divers have been following a similar procedure for quite a while, although they leave out the intermediate stage because chambers aren't available to them. I know quite a few who breathe oxygen in the water at the 20- and 10-fsw stops as an extra safety precaution when using Navy Tables, as a way of making up for the Tables' lack of conservatism.

Whenever I've breathed oxygen in the water during deco, I've always *felt* so much better after the dive - more energetic and somehow cleaner, as if my body had been purged of those nasty inert gas molecules that have an unpleasant tendency to bubble. In truth, the body *has* been purged of some of the nitrogen absorbed during the bottom phase of the dive - more than would have been eliminated during an ordinary air decompression.

When little or no nitrogen is taken into the body during deco, the rate of its elimination is increased.

Here's an example for comparative purposes in order to demonstrate the concept. According to Navy Tables, a one hour dive on air to 130 fsw requires 86 minutes of decompression if air is breathed as the deco gas. But if oxygen is breathed during decompression, the total ascent time is reduced to 53 minutes - a savings of 33 minutes. The hang on oxygen is less than two-thirds as long as the hang on air.

Blood Brothers: Nitrox and Oxygen

Take for example the *Monitor* trip I described in the chapter on "Submersible Air Decompression Computers." The required decompression for a 25 minute dive to 230 fsw was 165 minutes. The very next day I conducted the same bottom profile, but got out of the water after only 95 minutes of decompression (and lived to write about

it). I saved myself 70 minutes of hang time. How?

I breathed different blends of nitrox at various stages of the ascent, and breathed pure oxygen at the last two stops.

Here's what I did. I carried two sling bottles, one filled with nitrox-36 and one filled with nitrox-51. During my ascent I switched from air to nitrox-36 at a depth of 90 fsw. I decompressed for specified times and at succeeding 10-fsw stops until I reached the 50-fsw stop, at which point I switched to nitrox-51. I breathed nitrox-51 while ascending to the 20-fsw stop, at which depth I completed my decompression on surface supplied oxygen.

The advantages of such a regimen are obvious. So are the complications: three gas switches requiring a mass of equipment for all the delivery systems.

Consider this profile produced for me by John Crea. After a 20 minute dive to 250 fsw I ascended at the rate of 30 feet per minute to my first stop at 90 feet, where I switched to nitrox-36 for two minutes. I continued to breathe nitrox-36 for each successive stop until I reached 20 fsw, where I switched to surface supplied oxygen for the last two stops. Total deco time was 61 minutes.

This may not sound attractive when compared to the 59-minute penalty sentenced by Navy Tables, but then I've already put those Tables in their place.

Let it be known that when I got out of the water (with great trepidation) my wet computer, based on the Buhlmann algorithm, wanted me to hang for another forty-five minutes. I ignored it. The computer couldn't handle the situation; it pouted for a whole day.

Billy Deans goes a step further by pulling all his oxygen decompression at the 20-fsw stop, where the pressure differential forces more nitrogen out of solution than at the 10-fsw stop. He does not do this to reduce decompression time but simply to increase the rate of nitrogen elimination.

However, when pushing the ppO_2 to the limit, you must be careful not to max out the CNS clock, where the risk of toxicity is increased (see section below).

Decompression Efficiency

In the immortal words of John Crea, "accelerated decompression is not faster, it is more efficient." By decreasing the partial pressure of nitrogen in the gases breathed during deco, the nitrogen in solution can be eliminated at a more effective rate. That this efficiency is also faster is just a coincidental foible in our favor.

Deco Gas in Parlance and Practice

Any gas breathed during the decompression phase of a dive is called a deco gas. This gas may be nothing more than air, it may be oxygen, it may be any blend of nitrox, or it may be a combination or all of the above.

Two different nitrox blends were used in sequence in the examples just given because the second mix contained a lower percentage of nitrogen than the previous one. Reduced percentage of nitrogen equates to reduced partial pressure, which optimizes the efficiency of nitrogen elimination.

The higher percentage of oxygen in a deco gas has nothing to do with decompression. Quite the contrary, it inhibits the use of nitrox blends which contain even lower percentages of nitrogen, because of the risk of inviting oxygen toxicity.

If you wanted to make accelerated decompressions with the most efficient use of deco gases, how would you go about it? Technically, you could reach near optimum decompression efficiency by breathing a different nitrox blend at every 10-fsw stop. Each successive blend would contain a lower percentage of nitrogen (and a higher percentage of oxygen) than the previous blend.

The precise mixture of each deco gas would be dependent upon the highest allowable partial pressure of oxygen (1.6 ATA at rest). However, you would then have to consider the ticking of the CNS clock (see section below).

In other words, at 130 fsw you would switch from air to nitrox-32; at 120 fsw you would switch to nitrox-34; at 110-fsw you would switch to nitrox-36; and so on. This procedure would get you out of the water in an incredibly short time. But lugging all those tanks around might be logistically impossible, unless you were able to stage them at the appropriate depths (say, along the descending tunnel of a cave) or have them staged for you in advance.

Thus there's a point at which decompression efficiency becomes impractical or uneconomical,

and you have to reach a compromise. Cave divers have more latitude in this regard because submerged tunnel systems boast a controlled and predictable environment. In the open ocean, where the whims of weather and the vagaries of waves and current are inconstant, the measure of control is considerably reduced.

Accelerated Decompression Tables

The profiles I described above are proprietary. They were generated by specially developed programs from the computers of Bill Hamilton and John Crea. Others are available.

These programs take into account not only the maximum allowable partial pressure of oxygen, but also the cumulative exposure to oxygen under pressure. Total exposure is tracked by assigning a percentage of the CNS clock for each depth and duration, then plugging the accumulation of percentage points (the build-up of oxygen) into the equation so the recommended limit is never exceeded. (OTU's will always fall within the accepted limit as long as the CNS clock has not been exceeded.)

This is necessary because even though the accepted standard ppO2 is maintained, high-pressure oxygen exhibits a cumulative effect as well. If either the ppO2 or CNS clock are exceeded for any length of time, oxygen toxicity may result.

Oxygen Related Maladies

Commonly called oxtox, oxygen toxicity is high-pressure oxygen poisoning. There are three kinds.

Pulmonary or lung oxygen toxicity affects only the lungs and is about as serious as a bad cold. Whole-body or somatic oxygen toxicity affects the body in general with mild symptoms such as nausea, dizziness, headache, numbness in fingertips and toes, and reduced aerobic capacity. Both of these forms of oxygen toxicity occur only after exposures to high partial pressures of oxygen for half a day or more, and are therefore likely to affect divers in habitats or those making endurance or saturation dives rather than the typical technical diver.

The onset of pulmonary and whole-body oxtox is estimated by establishing a threshold accumulation of OTU's beyond which symptoms will probably occur. OTU stands for "oxygen toxicity unit" or "oxygen tolerance unit." The unit is calculated by an equation that correlates length of exposure to partial pressure of oxygen.

The third kind is central nervous system oxygen toxicity, or CNS oxtox, and is of greater concern to technical divers because it can be brought about by making dives of relatively short duration during which high partial pressures of oxygen are breathed. CNS oxtox affects the brain and can lead to seizures and unconsciousness.

Despite all the hoopla about divers breathing pure oxygen in the water (more scientist scare tactics), I leaven all objections by noting that the U.S. Navy promotes the use of oxygen rebreathers on *working* dives to a depth of 25 fsw for as long as 75 minutes.

Reduce the depth and the allowable exposure is extended: 110 minutes at 20 fsw, 150 minutes at 15 fsw, 240 minutes at 10 fsw. Even if you breathed oxygen only at the 10-fsw stop, it is unlikely that you'd require four hours of decompression. (If you stayed down that long you've got other problems.)

The question then arises: should you be worried about oxygen toxicity for decompression purposes? Worried, no. Forewarned, yes.

Like so many other susceptibilities to gases at increased partial pressure, tolerances vary from person to person, from day to day, and from dive to dive. A CNS oxygen "hit" can prove fatal under water. Not directly because of the affect of oxygen on the brain, but because the loss of consciousness during convulsions can cause a person to drown.

Consider a Navy diver breathing oxygen in a chamber pressurized to a depth of 40 fsw. If he experiences any adverse symptoms he can simply let go of the oxygen mask and breathe the air in the chamber. For that reason, the oxygen mask is not strapped in place around the head, but is held in the diver's hand.

Likewise, a Navy diver breathing oxygen that is fed to him from a rebreather is at very little risk because the system includes a helmet or full-face mask. A seizure or loss of consciousness won't necessarily result in death.

What can a decompressing technical diver do to avert such problems? First of all, don't push the CNS clock. Maintain a ppO₂ that is lower than the accepted standard limit: say, 1.4 ATA. This offers leeway for extended duration.

It is further recommended that you take an "air break" every twenty or thirty minutes. This isn't like the air brakes in olden automobiles: brakes that were operated by compressed air. In this instance an air break is more like a coffee break. It's an interruption in the decompression profile: one that allows you to get away from oxygen and back on air. Take five, as they say on the job.

The time you spend on air counts as "bad" time. It's not part of the profile, so you can't add it to your countdown time. Again, think of it as a coffee break: time during which no useful work is accomplished. The difference is that you get paid for coffee breaks, whereas you have to pay for air breaks by prolonging your time in the water.

It goes without saying that if you feel any symptoms of oxygen toxicity, spit out the regulator and go back to air. In that sense, oxtox can be compared to its opposite number, mountain sickness (which is a *lack* of oxygen reaching the brain). Get off the mountain and it goes away, stay there and it won't; it might get worse.

Symptoms of Oxtox

What is oxygen toxicity and what does it feel like? Well, no one really knows what it is, only what causes it and how it affects the body. As I stated in the previous section, oxtox is caused by too high a partial pressure of oxygen for too long a time: a function of depth and duration.

According to the highest accepted limit of 1.6 ATA (without the time constraint) ordinary air can become toxic at 218 fsw. That is the reason that most doctors of hyperbaric medicine prescribe such a depth as the limit for diving on air - not because of the bane of nitrogen narcosis.

But oxtox is also a function of the length of time one is exposed to high partial pressures of oxygen. This explains the sliding scale on which the Navy bases the use of oxygen rebreathers. So where does that leave us with respect to breathing oxygen during decompression? Pretty well in the clear as long as the durations are short and the ppO₂ is kept low - 1.4 ATA is considered by most doctors as absolutely safe.

But - symptoms of toxicity can occur. Once when decompressing with my friend Rick Jaszyn, he felt a headache coming on at the 10-fsw stop. He thought it might be due to toxicity, so he switched back to air. In a couple of minutes the headache was gone.

Preliminary symptoms are usually manifested preceding a full-blown attack. Recognizing these symptoms will enable you to take evasive action should anything untoward occur. Initial indications might be muscular twitching of the face and lips, tingling sensations in the fingertips, nausea, dizziness, tunnel vision, and ringing in the ears. Other symptoms closer to the onset of total toxicity seizure are fatigue, confusion, anxiety, and uncoordinated or random movements. The final stage is convulsion and unconsciousness.

Mountain climbers know that when altitude sickness occurs, the only cure is to get off the mountain. By the same token, if a diver suspects that he's nearing the limit of his tolerance to oxygen or feels the symptoms of toxicity beginning, his only recourse is to reduce the partial pressure of oxygen he is breathing. This can be accomplished either by switching to a gas whose partial pressure of oxygen is lower, or by ascending to a shallower depth.

Seizure from Oxtox

Some doctors say it is possible to go straight into seizure without warning. Think about that before deciding to extend your oxygen decompression to the 30-fsw stop.

Dr. James Clark, of the Institute for Environmental Medicine, wrote that the "risk of CNS toxicity is considerably increased during conditions that cause CO_2 buildup; possible causes of this could be suppression of hyperventilation during exercise, and the increased work of breathing dense gases. Some individuals have a reduced ventilatory response to exercise. Intermittent exposure improves oxygen tolerance." (From *Workshop on Enriched Air Nitrox Diving*, 1989.)

In other words, working too hard and not breathing fast enough. Reduce the dosage of oxygen by rotating it with air. If you feel any unusual symptoms while breathing oxygen in the water, stop breathing it. (But keep breathing - air.)

- Exercise decreases oxygen tolerance.
- Oxygen tolerance is decreased by water that is either too cold (less than 50°F) or too warm (more than 90°F).
- Tolerance to oxygen toxicity can vary daily and individually.
- Oxygen dose is cumulative.

It is important to understand the ramifications of the latter concept. If you've been diving deep then your body is already stoked with a large measure of oxygen. In that case, breathing oxygen during decompression at a partial pressure of 1.6 ATA may put you over the edge; that is, beyond the limit of exposure to where the risk of oxtox is high. Cutting back the partial pressure of oxygen is recommended, perhaps to as low as 1.3 ATA if the dive is long and you plan to work hard.

Put another way: because the duration at depth on most technical dives is short, a diver is more at risk during decompression because his CNS clock has been ticking throughout the dive.

A seizure induced by oxygen toxicity is no more harmful to the brain or body than an idiopathic seizure or an epileptic fit. The danger to life is drowning while in the grip of an attack. If an alert buddy brings a convulsing diver slowly to the surface, the patient will breathe normally after the seizure terminates - but both people incur the risk of DCI.

A diver wearing a full-face mask will continue to breathe despite paroxysms that may haltingly close the throat.

The jaw relaxes during unconsciousness. A diver drowns when he blacks out in the water because the regulator falls out of his mouth. But a convulsing diver clenches his jaw, thus gripping the mouthpiece firmly. It has been suggested that a strap that holds the regulator tight against the mouth may prevent the regulator from ever dropping out, although water might still be drawn in around slackened lips.

The CNS Clock

On most technical dives, the threat of CNS toxicity has become paramount long before the accumulation of oxygen toxicity units approaches the threshold where pulmonary or whole-body toxicity threaten. To phrase it differently, whereas pulmonary and whole-body toxicity are long-term affects of continued oxygen exposure, CNS toxicity can occur after short durations. For that reason great emphasis is placed on the oxygen dose: that is, the amount of oxygen absorbed by the body.

Instead of assigning units such as oxygen toxicity units, the theoretical threshold for CNS oxtox is measured by the accumulation of percentage points, with 100% being the maximum percentage allowed before the possible onset of convulsion. Every partial pressure of oxygen has been given a duration for exposure. Thus the rationale for the 45 minute time limit for breathing a gas in which the partial pressure of oxygen is 1.6 ATA, and the consequent sliding scale of increased durations for lower ppO₂'s.

This is also the rationale for using 1.4 ATA of oxygen as the basis for "safe" exposure - at least for the durations commonly encountered in technical diving. Lower ppO₂'s are recommended for longer missions or for a series of close repetitive dives. Thus cave divers making extraordinary penetrations and breathing from a string of stage bottles might calculate their gas mixes based upon 1.3 ATA of oxygen or less.

When we speak of running the CNS clock we are speaking in terms of the percentage of allowable oxygen dose that is being absorbed by the body at any particular moment during a dive. As the partial pressure of inspired oxygen is reduced (say, by ascending) the accumulation of percentage points is also reduced. In fact, the clock may come to a complete standstill and not start ticking again until the diver switches to a deco gas whose ppO₂ is higher than the ppO₂ in his bottom mix.

Ironically, a diver might be more at risk decompressing on pure oxygen at 20 fsw than he was on the bottom breathing air, partly due to the fact that his body has already absorbed a certain percentage of allowable oxygen dose.

In actuality, CNS oxtox is easy to avoid by simply maintaining a low partial pressure of oxygen in the breathing mix and decompression gases. When a diver who is decompressing on

oxygen at 20 fsw ascends to 10 fsw, he significantly reduces the risk of oxygen toxicity.

On the other hand, a diver who accidentally breathes oxygen at 200 feet (say, from a deco bottle) make take only a couple of breaths before going into convulsions.

Now that I've got all the warnings off my chest, let's put oxygen to use - not just as an added safety factor, but to get out of the water in less time than it takes to decompress on air.

Oxygen for Singular Decompression

Deco gas can be made available in a variety of ways and with a minimum of imposition. For intermediate depths and durations requiring a first stop in the 50-fsw range, a single tank of nitrox may be all that's necessary. As mentioned above, a cave diver can drop off his deco bottle as he proceeds through a cave, then reclaim it on the way out. In the ocean, a tank of nitrox can be suspended from the boat or anchor line; that relieves the diver of having to carry it with him throughout the dive.

More common as a single deco gas is oxygen. The beauty of oxygen is its simplicity: it doesn't have to be mixed, it is relatively inexpensive, demurrage on the cylinders is reasonable, and availability is universal. Furthermore, it is not difficult for an individual or a few friends to set up a simple cascade system for recharging deco bottles from storage cylinders. Such a system can be sequestered in the corner of a garage or basement. This will be discussed in detail in the chapter on "Expedition Diving."

GAS COMPARISONS

(130 fsw for 50 minutes)

Bottom Gas	Deco Gas	Decompression	CNS clock
Air	Air	81 minutes	17%
Nitrox-30	Nitrox-30	51 minutes	44%
Nitrox-30	Nitrox-70	32 minutes	57%

Note the significant reduction in deco penalty in the two nitrox scenarios compared to air. Breathing nitrox-30 on the bottom and during decompression saves the diver half an hour of hang time. By carrying a small sling bottle and breathing nitrox-70 for deco, an additional 19 minutes can be shaved off the hang, for a total saving of 49 minutes or 60%. With respect to oxygen toxicity, all three exposures fall well within the allowable 100% running time of the CNS clock.

Nitrox-70 - the Optimizer

An alternative to pure oxygen is a high nitrox blend. Ted Green favors nitrox-70 for several reasons. You can start breathing it from 40 fsw, the depth at which the partial pressure of oxygen in this mix reaches the maximum recommended limit of 1.6 ATA. By doing so, the total time of decompression closely approximates the time it takes to decompress on pure oxygen when breathed from 20 fsw. The benefit is in the reduced partial pressure of oxygen at subsequent stops. Nitrox-80 is another favorite compromise.

Oxygen shouldn't be pressurized higher than 2,400 psi. This means that only low-pressure sling bottles can be completely filled with oxygen. High-pressure sling bottles receive only a partial fill that is slightly more than half. But nitrox-70 can be delivered at 3,500 psi. Start with 2,000 psi of oxygen, and top with air to 3,500 psi.

DECO GAS TIME COMPARISONS

Breathing air at:	120 fsw/40 min.	150 fsw/40 min.	180 fsw/40 min.	210 fsw/40 min.
Deco Gas	Deco/CNS	Deco/CNS	Deco/CNS	Deco/CNS
Air 21% (all the way)	43 min./13%	74 min./19%	134 min./29%	195 min./37%
Nitrox-30 (from first stop)	37 min./15%	64 min./23%	108 min./35%	151 min./45%
Nitrox-40 (from 100 fsw)	32 min./16%	55 min./27%	88 min./42%	123 min./55%
Nitrox-50 (from 70 fsw)	28 min./22%	49 min./34%	74 min./61%	108 min./72%
Nitrox-60 (from 50 fsw)	26 min./22%	44 min./37%	69 min./52%	104 min./72%
Nitrox-70 (from 40 fsw)	24 min./24%	42 min./36%	67 min./52%	102 min./70%
Nitrox-80 (from 30 fsw)	22 min./25%	44 min./36%	70 min./53%	105 min./69%
Nitrox-90 (from 20 fsw)	25 min./22%	47 min./34%	78 min./51%	116 min./66%
Oxygen 100% (from 20 fsw)	24 min./33%	45 min./49%	75 min./72%	112 min./91%

The table above is intended for comparison purposes only and should not be used in actual practice. Air was used as a breathing gas during descent, on the bottom, and during ascent to the stop at which the recommended gas switch was made. The times refer to total stop time, not run time. The percentages refer to the fraction of the CNS clock that was run for the entire dive, not just for the decompression portion. All deco stops were calculated with a ppO2 of 1.6 ATA or less. If required stops exceeded a ppO2 of 1.6 ATA, those stops were conducted on air. All information was provided by Ted Green.

As the table amply demonstrates, air is the least efficient deco gas. Oxygen, while the most easily obtainable and logistically simple deco gas, is not the most efficient. Furthermore, breathing oxygen for deco runs the CNS clock to the highest percentage in every case.

I have underlined the most efficient oxygen enriched nitrox blends for each depth profile. Nitrox-80 is the best choice for a 40 minute dive to 120 fsw. Nitrox-70 is the best choice for the three deeper profiles, and differs insignificantly from straddling nitrox blends with respect to CNS clock time.

Over-the-Counter Oxygen

Don't let anyone convince you that you need a prescription to purchase oxygen from a commercial gas supplier. You don't.

Most divers don't even buy medical grade oxygen (or its more expensive relative, aviation grade). They settle instead for welding oxygen. Some people like to believe that there's a great dif-

ference in purity between grades, but the truth of the matter is that oxygen is oxygen is oxygen.

The main differences between medical grade oxygen and welding oxygen are that medical grade oxygen is analyzed and assigned lot numbers for tracking purposes, and that returned cylinders are "blown down" and "vacuum packed." That is, the cylinders are evacuated of all

Actual fsw=	10	15	20	30	40	50	60	70	80	90	100
Nitrox-32	5	9	13	22	30	39	48	56	65	73	82
Nitrox-36	2	6	10	19	27	35	43	51	59	67	75
Nitrox-40	0	4	8	15	23	31	38	46	53	61	69
Nitrox-45	-3	1	4	11	18	25	32	39	46		
Nitrox-50	-5	-2	1	7	14	20	26	33			
Nitrox-60	-11	-8	-6	-1	4	10					
Nitrox-70	-16	-14	-12	-9	-5						
Nitrox-80	-22	-20	-19	-17							
Nitrox-90	-27	-26	-26								
Oxygen	-33	-33	-33								

COMPARATIVE EAD FOR DECO GASES

In the left column are listed the most common nitrox deco blends. Across the top row are shown actual depths in feet of sea water. To determine equivalent air depths for a particular nitrox blend, match the chosen blend with the actual depth at which you plan to breathe that blend for decompression. In producing these values from the EAD formula, calculated depths were rounded up to the next whole number.

The purpose of this table is to demonstrate graphically the advantage of breathing nitrox blends containing increasingly higher fractions of oxygen. Nitrox blends in which the percentage of oxygen is less than 32% do not yield effective gains in the reduction of decompression in comparison to air.

Negative values signify a lower partial pressure of nitrogen in the mix at depth than exists in air on the surface. For decompression purposes, the more negative the number, the better the purging effect because nitrogen is eliminated faster. By extrapolation, pure oxygen offers the least amount of resistance to expired nitrogen because expired nitrogen is not "bucked" by inspired nitrogen.

Ironically, you can eliminate more nitrogen underwater by breathing a nitrox blend with a negative number than you can by breathing air during surface interval. This is due to the absence of inspired nitrogen pressure.

Warning: in two instances the values shown exceed 1.6 ATA , which is the recommended safe limit for the partial pressure of oxygen with respect to oxygen toxicity. These borderline instances occur for nitrox-40 at 100 fsw (for which the actual depth is 99 fsw) and for oxygen at 20 fsw (for which the actual depth is 19.8 fsw). I have extended the rows in these instances because these deco gases are generally breathed at these depths. However, the prudent diver should keep in mind the fact that pushing the limits increases the risk factor.

This table was created with calculations made from the following formula (all values were rounded up to the next whole number):

$$EAD = \frac{[(1.0 - fO_2) \times (D + 33)] - 33}{.79}$$

remaining oxygen and then refilled from scratch from a new lot. The lot is analyzed and assigned a tracking number.

Returned cylinders of welding oxygen are simply topped off with more oxygen. If a tank is emptied for any reason (say, to replace the valve), oxygen is pumped into the cylinder without the cylinder first being evacuated. In that case, traces of atmospheric air in the cylinder can be construed as an impurity.

Some alarmists plead that welding oxygen cylinders can be contaminated by sloppy welders. Check valves in a welding rig prevent gas from backing into the cylinder. However, if a welder fails to install check valves, then empties the oxygen cylinder during a burn, acetylene can be sucked up the hose into the oxygen cylinder. Although this possibility is remote, you should keep the thought in mind when deciding on which grade oxygen to buy.

It is true that *some* commercial gas suppliers will not sell medical grade oxygen without a prescription. Then, if they suspect that the inquirer intends to breathe the product, they refuse to sell him oxygen of any grade. This seemingly nonsensical attitude is unrelated to safety considerations. It is a reaction to the seller's presumed liability.

Most technical divers sidestep the issue altogether. They feel out the gas supplier anonymously. Then, if they think that problems might arise from full disclosure, they don't let the supplier know that the oxygen is for blending or for decompression purposes. What the supplier doesn't know can't hurt him.

The Deco Pony Bottle

With the widespread employment of oversized tanks equipped with double valve manifolds and isolator valves, some divers, who were used to carrying an alternate air supply for emergency situations, now use their pony bottles for deco gas. This is an excellent idea as long as you don't breathe oxygen or nitrox accidentally at depth, where the partial pressure of oxygen could prove immediately fatal.

Some changes in rigging are suggested:

Nitrox-	32	36	40	45	50	60	70	80	90	O2
1.6 (45 min)	132	113	99	84	72	54	42	33	25	19
1.5 (120 min)	121	104	90	76	66	49	37	28	21	16
1.4 (150 min)	111	95	82	69	59	43	33	24	18	13
1.3 (180 min)	101	86	74	62	52	38	28	20	14	9

ppO2 DECO LIMITS

This table is similar to the one printed in the nitrox chapter (ppO2 Depth Limits).

The left column lists the partial pressures of oxygen in atmospheres absolute, with the total recommended single exposure given in parentheses. The top row shows the percentages of oxygen in the most common nitrox decompression blends, with pure oxygen shown in the right column. Match the desired partial pressure of oxygen with a particular nitrox blend in order to determine the maximum depth in feet of sea water at which the blend should be breathed.

To use these values in conjunction with decompression tables, the numbers should be rounded down to the next lower increment of 10. For example, nitrox-32 at 1.6 ATA yields a value of 132 fsw, which is rounded down to 130 fsw.

This table was created with calculations made from the following formula: Depth = ppO2/fO2 - 1 ATA. For example, a ppO2 of 1.6 ATA divided by .4 (the fraction of oxygen in nitrox-40) yields 4 ATA; subtract 1 ATA and the result is 3 ATA, or 99 fsw. (1 ATA equals 33 fsw.)

Warning: pure oxygen is commonly breathed at 20 fsw for decompression purposes. But note that 1.6 ATA of oxygen is shown as 19 fsw. By calculation the limit is actually 19.8 fsw - a decimal which is extremely close to 20 but which is rounded down to 19. This demonstrates how close oxygen comes to the recommended limitation with respect to CNS toxicity. On the up side, decompression is generally conducted in a condition of rest.

• Instead of routing the regulator hose over the shoulder, route it under an arm where it isn't readily accessible. This way you can't grab it by mistake.

• Secure the pony bottle upside down and route the regulator hose past your waist. An extra long hose is necessary, as well as some way of securing the second-stage so it doesn't dangle out of reach.

• Place a rubber band across the mouthpiece as a reminder that the regulator is for decompression only. Your tongue will notice the difference.

• Replace the rubber mouthpiece with one that is flavored.

• If you are extremely agile, don't turn on the deco bottle until you need it.

In case you skipped over the section dealing with oxygen cleaned tanks in the nitrox chapter, let me state here emphatically that all scuba equipment intended for use in a pure oxygen environment must be decontaminated for oxygen service. Replace the soft parts of valves and regulators with oxygen compatible materials, and lubricate only with greases that are not petroleum based. Read the section below on "Oxygen Cleaning Regulators."

In-water Oxygen Rigs

On most deep air diving trips accelerated decompression has become the norm. This is accomplished with simplicity by breathing pure oxygen during the final two stops, where the partial pressure lies within the accepted safe limit.

Oxygen can be carried in a sling bottle or pony bottle, but for large-scale boat operations, where too many accessory tanks can clutter an already crowded deck, an in-water rig can best serve the needs of the multitude.

Depending upon the number of divers and the length of the trip, one or more large storage bottles of oxygen can be lashed down securely to the deck, perhaps out of the way under a bench or dressing station, or stowed upright and lashed to a rail or bulkhead. Once the grapnel has been set in the wreck, the rig is connected to one of the bottles and then lowered into the water.

Designs vary. The basic rig consists of a pressure gauge on the oxygen cylinder, a scuba thread adaptor, one regulator first-stage, several long hoses (called "whips") each with a second-stage, and a stout vertical line held down by heavy ballast.

The ballast line should dangle some fifty feet into the water. This is so divers can transfer to it from the anchor line in order to conduct their deeper stops. This relieves thronging on the anchor line. The oxygen regulators, however, should not reach down much deeper than 20 fsw. A few feet leeway is okay to allow for scope, but you don't want people drifting down too deep while breathing oxygen.

HP versus LP

There are a number of ways of rigging oxygen regulators for deployment, with variations for each. Basically these are split into the high-pressure system and the low-pressure system (the pressure ratings being relative to each other). Let's begin with the high-pressure system.

Screw an oxygen nut and nipple (CGA 540) into the oxygen cylinder. Attach a T fitting to the nut and nipple. One of the T's ports can be used for an oxygen pressure gauge (not a submersible pressure gauge). Screw a thread adaptor into the other port and screw one end of the high-pressure hose into the adaptor.

The high-pressure hose should be thirty or forty feet long. This enables you to snake the hose across the deck from the oxygen cylinder to wherever it's convenient to slip the hose over the side. At the other end of the high-pressure hose connect a regulator first-stage, and lower it into the water to a depth of 20 fsw. The low-pressure ports are used for second-stage whips.

There are disadvantages in this system to consider. For one, high-pressure hose is stiff and difficult to handle, even when it's not pressurized.

Uncoiling the hose can be a frustrating exercise. It wants to retain its coils, so when the hose is strung across the deck, it rises in a series of loops that constantly snare people's feet. Equally as onerous is coiling the hose for stowage.

Another handicap is having hose that is under high pressure in the water, and having the first-stage so far away from the cut-off valve. If the first-stage springs a leak or blows out under water, an awful lot of oxygen can be lost in the time it takes for topside personnel to notice the eruption of bubbles breaking the surface and to close the valve. (That the gas escaping is oxygen has no special significance.)

In the low-pressure system, a regulator is connected directly to the oxygen cylinder. This regulator can be one of two kinds: an oxygen regulator or a scuba first-stage. A scuba thread adaptor is still necessary, either between the oxygen regulator and the low-pressure hose, or between the tank valve and the scuba first-stage.

An oxygen regulator of the kind used in welding is comparatively inexpensive. The unit is equipped with two gauges: one for tank pressure and one for intermediate pressure. These gauges are large and easy to read. The intermediate pressure can be adjusted in a trice with a turn of the handle.

Welding regulators have several drawbacks that make them less than ideal for use with in-water oxygen rigs. Most common models cannot achieve the required working pressure to drive scuba second stages: they work in the range of 60 psi. Models exist that can operate at 120 psi to 140 psi, but they cost more.

Furthermore, welding regulators do not find the briny environment friendly. The regulator housing and the thin metal gauge casings are prone to rust. Care must be exercised to protect the assembly from spray and resulting salt deposits, and it should be rinsed thoroughly to prevent the build-up of corrosion. Because the regulator housing and gauge casings are not watertight, leakage during the rinsing process can eventually damage the internal mechanisms.

Scuba first-stages demand far less attention since they are designed for salt water service. In case of a mechanical malfunction, the regulator can be replaced on site with any spare regulator that isn't being used. The inoperative unit can later be serviced readily at most reputable dive shops. A submersible pressure gauge screwed into the high-pressure port will enable topside personnel to monitor the tank pressure.

Once either one of these regulators is placed on the oxygen cylinder, the low-pressure hose is fed into the water. Low-pressure hose is as pliable as rubber garden hose and is just as easy to handle. At the opposite end of the hose goes a block. (This is where the first-stage is located in a high-pressure system.) The block should be suspended no deeper than 20 fsw. It has three outlets for the connection of second-stage whips.

One variation of the low-pressure system does away with the block altogether. Instead, the whips are fitted directly to the first-stage on the oxygen cylinder. Each whip has to be long enough to reach the required depth. The virtue of this system is that decompressing divers have such long lengths of hose that they can gambol about under water and get away from the maddening crowd. The only deficiency is the loss of a shut-off valve.

Rigging for Success

To ensure that the first-stage or block is lowered to the pre-selected depth (generally between 10 and 20 fsw), wrap colored tape at that distance from the end, and let that be the waterline mark. Then you can see from the surface that the proper length has been deployed under water. Snug the hose accordingly.

It is less restraining on decompressing divers if the second-stage whips are longer than the customary length. At least seven feet is recommended, although ten feet or longer is more comfortable. Short whips force the divers to cluster together. Long whips permit them to get away from each other giving each his personal space, and makes possible the use of jonlines in strong current or rough seas.

Also recommended are shut-off valves at the block: one for each regulator. The mouthpieces have a tendency to trickle as the vertical line rises and falls with the swells and the second-stages are jerked up and down. A second-stage that is out of tune will also bleed bubbles. The resultant loss of oxygen is annoying.

Worse yet, usually once an oxygen rig is deployed, the system is kept operational for the rest of the day in order to accommodate divers whenever they make their dives. Over a long period of time the loss of oxygen can be significant.

Moreover, if one second-stage suffers a catastrophic free-flow, the oxygen tank will quickly drain unless the free-flowing mouthpiece is isolated from the system. In practice, individual shutoff valves are kept closed when not in use. The diver opens the valve to breathe oxygen, then shuts it when he's done.

The simplest valve is the quarter-turn bell valve. This kind has a handle which opens and shuts the gate with a flick of the wrist. The valve is closed when the handle is positioned perpendicular to the line of flow; it is open when the handle is parallel with the line of flow. Despite the ease of operation, this type of valve is more maintenance intensive because of its tendency to leak.

More secure is the screw-down valve. This kind has a knurled knob that is turned clockwise to close the gate, counterclockwise to open it. (Righty tighty, lefty loosey.) These valves are dependable and they seldom leak.

Rigging the Vertical Line

Boat captains have gotten creative with the way they set up their oxygen rigs. Some rigs are hung off the stern, some off the side; it doesn't seem to make much difference.

Theoretically, once a boat is grappled it should align itself with the current so it pitches with the waves. This makes a stern rig bouncy. Throw theory out the window, however, for on too many occasions I've seen a crosswind overpower the current and give the boat a bad set, in which case it rolls unmercifully because the beam is shorter than its length. Here comes rock 'n roll.

Offering more accommodation than the single vertical line is the double. In this system, two vertical lines are suspended beneath the boat, each held down by its own ballast. The two lines are then connected at a depth of about 10 fsw by either another line, a length of PVC pipe, or an aluminum beam. This allows decompressing divers to spread out while giving them something to hold on to.

An important adjunct to the vertical line is a travel line or traverse line that is connected to the anchor line. This line enables returning divers to locate the vertical line in poor visibility and makes it possible for them to attain the line when a strong current is running crosswise. The travel line must be held at sufficient depth to permit returning divers to use it without rising above

their ceiling.

The current line can double as a traverse line if it connects the anchor line to the oxygen rig at sufficient depth. In that case, a diver entering the water can drop straight down the vertical line until it intersects with the rope connected to the anchor line, and pull himself against the current.

All ropes should be fat so they present a good grip. Use nylon; polypropylene is too slippery.

The vertical line needs to be fitted with rings, loops, or some such method of attachment for snap hooks. Several of these should be established at various depths. Because the slightest current will waft the whips out of reach, they need to be clipped in place along their length. By rigging the whips in this fashion (instead of having the whips taped to the line), you can unclip a second-stage when you want to breathe off it, then carry it up with you to a shallower depth.

Keeping the Rig Down

Ballast is very important - and it has to be very heavy. A spare weight belt will not fill the bill. The ballast should weigh in the neighborhood of fifty pounds or more. You can use scrap iron such as a small engine block, a bucket full of cement, a steel anchor (not aluminum), old chain or ship's chain links, and the like.

Lead works just as well but is more expensive. Stack a bunch of scuba weights so their slots coincide, then slip a long nylon webbing down through one side and up the other. Don't use a buckle to hold the ends of the webbing together; it won't hold under the strain. Tie the ends of the webbing in a knot, or cross loop them through a couple of weight belt retainers.

The rope should be thick so you can get a grip

on it, both for lowering and for raising. Heavy-duty gloves will make the rope easier to handle, and might save a few layers of skin. You can knot the rope around the ballast each time, but it's better to devise a system of securement in which the rope is looped and is connected by means of a carabiner or shackle.

Although the inclination is to toss the ballast overboard, don't do it. The shock of the weight reaching the end of its rope can break the line. Lower the ballast slowly into the water. And make sure the bitter end is tied to a cleat. More than one ballast block has gone to the bottom due to carelessness.

Why am I putting so much emphasis on ballast? Because without enough weight to hold it down, the line has a tendency to fly. You might not think that a current mild enough to permit diving would be strong enough to lift a weight and swing the line at an angle. But get a few divers hanging onto the line and their combined drag and buoyancy will raise all but the heaviest of loads.

You can scramble down deeper in order to maintain your ceiling, but because the whips have a preset length that is predicated on a vertical attitude, once the whips are soaring you can no longer reach the regulators. Then you have to decelerate your deco with the air on your back.

Even worse is the case in which the anchor line parts or the grapnel pulls out of the wreck. If the boat is driven by a hard wind, the vertical line will billow out like spider web in a storm. There's nothing more harrowing than trying to maintain the depth prescribed by your decompression profile, and finding yourself being dragged above it like a lure on a trolling line because of your need

to keep the oxygen regulator in your mouth.

Only one "hang from hell" is needed to convince you of my wisdom.

Switching Oxygen Cylinders

It's important for topside personnel to watch the gauges on the oxygen cylinder in use so they can switch the rig as soon as the tank gasps its last. This is done in the least amount of time, in the manner in which an Indianapolis pit crew changes tires on a racing car between laps.

It is customary to let decompressing divers know that a switch is about to take place. Sometimes this is accomplished by the simple expedient of shutting off the tank. I can tell you that this is effective. Sucking on a regulator that has suddenly ceased to deliver gets the message across emphatically - although it does come as a bit of a surprise. I hate it when they do that.

I like it better when they lower a slate with a written statement of intent. A good pit crew can switch regulators before decompressing divers have sucked out the oxygen in the hose, especially if the hose is long.

Better yet, two oxygen cylinders can be manifolded together by means of a T and a pigtail and a couple of check valves. This enables topside personnel to close one valve while opening another so there is never a gap in the flow of oxygen.

Stowage and Maintenance

An oxygen rig consists of a lot of rope and hoses. These can be coiled in a large plastic storage container such as those obtainable from most discount stores. They are usually found in the hardware department. Make sure to get one with a lid. They are cheap yet take a fair amount of abuse.

Do not use a milk crate. A storage container with solid walls and a lid will keep salt spray off the rig when it's not in use, and can double as a rinsing tub.

Treat an oxygen rig the same as you would your other gear. Soak or rinse it in fresh water then lay it out to dry.

Regulators should be serviced regularly. Check the mouthpieces often. They get torn or chewed up worse than those on regulators that you carry with you. This is because you're not physically connected to them, and you tend to bite down on them hard when they're about to be ripped out of your mouth by a rogue wave.

Ballast should be stowed separately. Otherwise, the container will be so heavy that you can't lift it. If you use scrap iron for ballast, keep it in a milk crate or a similar type of box that has grips or handles. That way you can haul it about without wearing gloves, and it won't leave gouges or rust stains in your vehicle.

Oxygen Cleaning Regulators

Regulators used for oxygen rigs should be cleaned for oxygen service. This is not so much because of possible explosive hazards, but because oxygen deteriorates rubber products more readily than air. Consequently, an uncleaned regulator will require more maintenance if it is used for breathing oxygen.

This doesn't mean that catastrophic failure is more likely to occur, but that annoying leaks will develop with increased frequency. Replacing rubber seats and o-rings with those made of viton will ensure that your regulators give lasting and reliable service.

A good technical dive shop should offer oxygen cleaning service. If you're a do-it-yourself person, you can purchase the parts and chemicals and clean your own regulators at home. Some regulator manufacturers sell kits that contain all the parts that need replacing for conversion to oxygen service. Instructions are included.

The procedure used to decontaminate regulators is much the same as that used for tank valves, described in the previous chapter. The cleaning procedure is not complicated, but disassembling and reassembling the regulator requires some basic knowledge of how regulators are constructed and put together.

Do not disassemble a regulator unless you know for certain how to put it back together.

My advice is to leave regulator cleaning to the experts and technicians who work on regulators every day. If you simply *have* to do it yourself, here's how. Only the first stage of the regulator needs to be cleaned.

Remove all stock soft parts such as seats, diaphragms, and o-rings. These are made of materials such as Teflon, neoprene, urethane, and buna-N; these materials are not compatible with oxygen service.

Soak the metal parts in heated vinegar to rid the surfaces of corrosion. This could take several hours, or you might soak them overnight. The use of an ultrasonic cleaner will hasten the process. When all signs of corrosion are gone, rinse off the vinegar with tap water and then soak the parts in DOT-113 (the same solution used to clean tank innards and valves). A twelve-hour soak should do the job. Rinse thoroughly with hot tap water and let dry.

Replace the soft parts with equivalent parts made of viton. Remember that all these parts look alike no matter what materials are used in their construction, so be careful not to mix them up after they are removed from their packages.

Lubricate all parts and threads with a medical oxygen lubricator such as Christo-lube or Krytox grease. Do *not* use silicone grease or any standard petroleum-based lubricant.

After reassembling a regulator on your own, let your mother-in-law try it out in the pool. The deep end.

As a general practice, the hose and second stage are not cleaned for oxygen service despite the fact that rubber (of which the hose is made) is not oxygen compatible. This is because oxygen flows through the hose and second stage at reduced pressure, at which the risk of ignition is negligible.

This same general practice holds true for submersible pressure gauges and their high pressure hoses. The pinhole in the hose connector that screws into the first stage is a flow restricter: not only does it prevent a catastrophic gas loss in the event the hose bursts, but it reduces the rate of oxygen flow to a point at which the risk of ignition is negligible. However, a high-pressure swivel between the hose and the first stage should be cleaned for oxygen service.

Some manufacturers sell regulators, hoses, and submersible pressure gauges that are cleaned at the factory for oxygen service. If you need to purchase extra gear in order to set up an oxygen system or to rig auxiliary tanks for oxygen deco, then you may as well buy everyday regulators and gauges that are already oxygen cleaned.

High-pressure whips are usually made of thermoplastic material, which is compatible with oxygen service.

A regulator cleaned for oxygen service should thereafter not be used on tanks which are not themselves oxygen cleaned and filled with clean-grade gas. Otherwise, contaminants might leave oily residues on regulator parts and assemblies, thus compromising the oxygen-cleaned integrity and necessitating recleaning.

Surfacing on Oxygen

There is strong evidence to support the claim that rigorous muscular action soon after a long decompression can induce DCI. This is because the surface is where the ambient pressure is the least and where bubbles are most likely to form.

For that reason, cave divers have developed the habit of continuing to breathe oxygen *as they ascend* to the surface from the final stop, and then floating for several minutes while still breathing oxygen before climbing out of the sink hole.

This procedure is difficult to implement on ocean dives, where hanging onto a line in a strong current or crashing waves might be just a rigorous as climbing the ladder onto the boat. At the very least, wreck divers should refrain from hustling tanks or pulling up the grapnel immediately after a dive.

Only When I'm Near You

Keep in mind that reliance on accelerated decompression profiles *demand* accessibility to the gas or gases specified by the program. You *must* be able to get back to your gas. The default option is to hang out your computer or back-up table - and to make sure that you have enough air to complete the longer decompression.

A deep dive followed by a non-accelerated decompression can be *significantly* longer. Recall my comparative examples with respect to the *Monitor*. If there is any possibility that your bottom gas supply might not suffice for a normal decompression, then you must carry your deco gas with you, or stage it where there is no chance of losing it.

Some divers carry a small bottle of oxygen as a back-up. They breathe off it only if they can't get back to the surface supply rig. This is good insurance.

Floating Deco, or Get my Drift?

For the 1990 *Monitor* expedition I developed a drift decompression system in order to deal with the predominantly strong currents that are so often encountered off the Diamond Shoals of North Carolina, where the Gulf Stream speeds by on its northeasterly journey from the Caribbean to the British Isles.

Boats are not permitted to grapple into the *Monitor* so as to protect the wreck from damage. But even had that not been the case a drift decompression system was warranted in consideration of fatigue: hanging onto a stationary line for a couple of hours against a forceful flow of water does not make for a healthy decompression.

For inspiration I capitalized on the emergency decompression procedure in which a diver drifts with the current under a floating liftbag - which I have had to do in my sordid past. I substituted a fishing float for the liftbag, hung weights beneath the float on a long vertical line that offered grip and depth control, and devised a "breakaway" line that a support diver could connect to the down line prior to the dive. The down line is supported on the surface by its own fishing float.

The down line must have a loop or ring somewhere between 50 fsw and 100 fsw. One end of the breakaway line is secured to the vertical line that hangs beneath its float. The other end has a loop or ring. A carabiner or snap hook connects the loop or ring of the down line to the loop or ring of the breakaway line.

The two floats bob on the surface with a discreet distance between them. The up-current float is secured to the down line, the down-current float is secured to the drift system. Divers are dropped into the water up-current of the down line. They drift back to the first float, grab onto the down line, and descend. After leaving the bottom they ascend the down line to the junction of the breakaway line, release the clip or carabiner, then drift along with the current.

All the tension of hanging onto a line is gone.

Divers and the drift system move at the same rate of speed, like a shopper on an escalator or a traveler on an airport slidewalk. The divers swim or pull themselves along the loose breakaway line to the vertical line beneath the float, then work their way up as their decompression schedule demands.

Surface personnel know that the divers have completed the transfer as soon as the distance between the two floats increases. The boat can then be maneuvered alongside the float. A rope or lasso is tossed around the float and it is secured to the side of the boat. At that point deco bottles or an in-water oxygen rig can be deployed.

Once the divers transfer to the boat's hanging decompression rig the float and attendant lines and weight can be brought aboard and stowed for future use.

Caution: problems may occur due to fast current, high wind, or a combination of both. A boat may be blown in a direction different from the set of the current. The higher the boat's sides and superstructure, the more surface area the wind has to push against. Extra heavy weights on the down line can help to alleviate the tendency of the system to sail at an angle.

Adrift over the Lusitania

A similar drift rig was employed on the *Lusitania* in 1994. The breakaway system was identical, but instead of just one float and a weighted vertical line we had five, in order to accommodate five teams of divers.

The breakaway line was connected to the vertical line hanging beneath the first float, which for the sake of clarity I will call position one. Position one was connected to position two by two carabiners and a five-foot length of PVC pipe. Position two was connected in the same fashion to position three. And so on to position five, from which an extra length of PVC pipe extended.

A large plastic slate was secured to the breakaway point on the down line. Each diver checked off his or her name going down and coming up. This way the last team knew for sure that everyone else was up and they could disengage the breakaway system without fear of stranding anyone on the down line.

The five positions were pulled taut by the current. But after the breakaway line was detached from the down line there was no strain to keep the positions in line. They tended to bunch up so that divers kept crashing into their neighbors. To avoid collisions, a support diver took the length of PVC pipe that extended beyond position five, swam it around in a circle, and connected it to position one. Crisscrossed lengths of PVC pipe kept opposite positions from coming together. The result looked like a square with a triangle protruding from one end, much like the home plate in a baseball diamond.

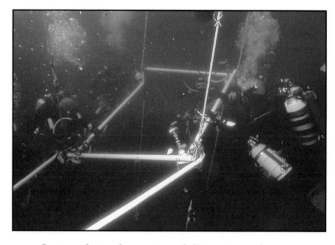

Secured to the vertical line at each position were two bottles of oxygen, already rigged with regulators. Each diver carried two sling bottles of nitrox for the deeper stops, then breathed oxygen at the last two stops. Metal loops spaced along the vertical lines enabled divers to unclip the oxygen bottles and move them up as the situation demanded. Extra long hoses made it possible to spread out: up or down the vertical line, along the PVC pipes, or to hover with neutral buoyancy.

Chief artificer of this system was Richard Tulley. The system was dubbed a "dec" station for the "dec" in "decompression." Dec is pronounced like "deck."

At no time was the boat connected to the dec station. The boat drifted nearby and collected divers as they surfaced after completing decompression. Once everyone was aboard, support divers disassembled the rig and pulled it onto the boat one position at a time.

Multiple Deco Gas Method

In order to circumvent reliance on surface supplied deco gases, technical divers have learned to carry two supplementary tanks with relative ease, sometimes three or more. The purpose is to put the SC back into technical SCUBA and dive completely self-contained.

If you've read the chapter on "Gear in Depth" then you're way ahead of the game. I've already described how accessory tanks can be carried with the least amount of hindrance. I don't need to repeat myself here.

I've also described in previous sections of this chapter how different blends of nitrox can economize on the time spent decompressing. For reasons of practicality, however, the most common multiple gas deco method is to carry but two auxiliary tanks: one with nitrox and one with oxygen.

Of the various nitrox blends, nitrox-32 has the advantage of being breathable at 130 fsw without exceeding 1.6 ATA of oxygen. The corollary of breathing deco gases from the 130 fsw stop is that it leaves you with that much more gas in your primary tanks. By carrying deco gases, you have in effect increased the amount of gas available to you on the bottom. Conversely, you could wear smaller tanks on your back if you didn't need the extra gas at depth.

It's a good idea mark the contents on an auxiliary tank. And not just the fact that the tank contains nitrox, but which blend of nitrox. If you grab the wrong tank while getting dressed, or breathe from the left side-slung instead of the right, your decompression could be seriously compromised. Place two large labels on each tank, one on either side: this will enable your buddy to confirm that you're breathing from the appropriate tank.

Some divers carry two side-slungs with the same deco gas in both. Each tank is therefore a back-up for the other in case the regulator malfunctions. Because they are breathing a single deco gas instead of two - one with a higher fraction of oxygen than the other - decompression will be longer, but it could be worth the peace of mind.

It is also possible to change regulators under water from one tank to another - and I know divers who have done that very thing in an emergency. Not that you can remove a regulator from a tank on your back, but it is relatively easy to remove regulators from side-slung bottles and to switch and replace them. Sure you'll get some water in the working parts and have to tear down the regulator afterward, but when you're facing a long hang in near freezing water, it might seem like the best alternative.

To dramatize the efficiency of the multiple gas deco method, compare these proprietary figures. A 45-minute dive to 130 fsw breathing air for decompression requires 77 minutes total ascent time. By breathing nitrox-28 as a deco gas, the total ascent time is reduced to 55 minutes. By breathing two deco gases - nitrox-28 and nitrox-70 - the total ascent time is further reduced to 32 minutes: less than half the decompression time than that required when breathing air, and a savings of three-quarters of an hour!

Enough said.

Accelerated Profile

Near the beginning of this chapter I mentioned proprietary tables. I highly recommend them. But suppose you find yourself on a trip where deco gases are made available unexpectedly (such as an emergency in-water oxygen rig) and you don't have your tables? What can you do to take advantage of the situation.

You'll hate me for saying this. Use Navy Tables.

I know I've gone to great lengths to emphasize that Navy Tables offer the least conservative decompression profiles in the history of the world, and should be used only if you have a recompression chamber within arm's reach. *But -* using Navy Tables while breathing oxygen at the 20- and 10-fsw stops is a viable alternative when proprietary tables aren't around. This is the *only* way I would use Navy Tables. Call it an acceptable option in an emergency.

In actuality, U.S. Navy Standard Air Decompression Tables coincide closely with computer generated nitrox or oxygen decompression tables. Why go the extra expense of proprietary tables, you may ask?

Because they're more accurate, because they're more reliable, and because they're based upon the latest scientific knowledge in decompression theory, such as slower ascent rates, deeper stops, more tissue compartments, and longer times for nitrogen elimination.

If you compare your dive tables with computer generated accelerated tables you can get a feel for how much time you save by breathing deco gases with higher percentages of oxygen than air. That way, if a situation arises in which oxygen is made available and you don't have any proprietary tables, you'll know how to shave stops and times off the table you ordinarily use.

As a general rule of thumb, you can't go wrong if you cut 25% off the printed decompression time.

Computer Back-ups and Hang-ups

Even if I don't follow my wet computer profile for decompression, I like to have a computer with me in case my accelerated deco goes awry. It's common practice not to open the valve on a deco tank until you're ready to breathe from it. That way your deco gas won't leak out during the dive. But once, when I opened the valve at my first deco stop, the regulator free-flowed uncontrollably. I had to decompress by computer. (I wasn't happy at all about the extra forty minutes I had to spend on the hang.)

Always have a back-up procedure available: sufficient air reserve, alternative air or deco gas supply, a computer or tables, and so on. An accelerated profile is an efficient way to decompress; it shouldn't be the only way.

The problem with most wet computers is that they get bent out of shape even if you don't. That is, they hang up for a day or so and can't be used in the mean time. That's how long it takes for the computer to reset itself. Models with a manual switch or a field-removable battery are preferred as back-ups.

Alternative tables and an extra timepiece work well for redundancy.

Even Navy Tables are better than nothing. Stop deeper, stay longer, and pray.

If your timepiece fails, practice counting from one thousand and one.

Programmable Wet Computer

The godsend for conducting accelerated decompression is a wet computer that can be programmed for deco gas switches. It's on the market now and it works like a charm. It's actually an advanced nitrox computer: the kind in which the nitrox blend can be selected before the dive.

Calling this state-of-the-art instrument a nitrox computer is like calling a DeLorean a passenger car, or the Space Shuttle a high-altitude airplane. It is so much more.

These models permit you to program one or two gas switches into the decompression profile. These deco gases can be two nitrox blends, oxygen, or a combination of nitrox and oxygen. All oxygen percentages and gas switches can be programmed in the field by means of external touch contacts prior to the dive. The program can *not* be changed once the computer is submerged.

It works like a sophisticated air computer, displaying run time, maximum depth, current depth, ceiling, stop times, total ascent time, and so on. During ascent you obey the ceiling and stop-time indicators as you would ordinarily. When you reach the stop at which you programmed the gas switch, the display flashes a reminder.

The computer does everything but switch regulators for you.

You continue to follow the ceiling and stop time indicators. At first it doesn't appear that the computer has gone into accelerated decompression mode. The countdown time counts down in the usual fashion - except that the internal clock mechanism has speeded up. The minutes go by faster than normal because they are less than sixty seconds long.

This is disturbing the first few times you use the computer. If you have doubts about its efficacy, compare it with a printed table. But remember that the computer will give you credit for multi-level excursions.

Do I need to mention once again that, despite the sophistication of the unit and my adoring rave revue, you still need a back-up? I thought not.

Dry Computer Option

In addition to proprietary computer programs that are generated and printed, there are programs that you can purchase outright and run on your own computer. Not to jump the gun, I will reserve further discussion in this regard for the following chapter. See the section on "Laptop Decompression Software" in the next chapter on "Mixed Gas Helium."

Is it Worth it?

Is accelerated decompression for you? This is a decision that only you can make. If you're uncomfortable with the additional risk factors that are entailed, then you may decide not to accelerate your decompression.

Another factor to consider is financial. While accelerated decompression decreases the time spent in the water, it increases the economic burden by necessitating the purchase of gear, gases, and decompression profiles that are not in the public domain.

From the comfort of your home it might not seem worth the investment in time and money, but think it over the next time you're turning into a prune while suffering through a long deco in cold water and rough seas, and wearing a leaky drysuit. You might change your mind.

MIXED GAS HELIUM
Alternative Breathing Media

While nitrox can be called the gas for the masses, helium is only for the few.

Concurrent with the mechanical improvements of underwater breathing apparatus are scientific advances in physiology and the study of the life-support requirements of the human body. The respiration of compressed air imposes decompression penalties and depth limitations due to many factors, most of which are not clearly understood.

Nevertheless, while scientists argue over the exact mechanisms responsible for nitrogen narcosis, oxygen toxicity, and decompression injury, experimenters have discovered ways of getting around the undesirable side effects that air under pressure inflicts upon the body and the brain.

This bold approach involves breathing gases and special gas mixes that are at variance with the atmosphere in which life on Earth evolved.

Deep diving has many shortcomings: the breathing supply decreases at an incredible rate, the surface is extremely far away, and decompression is a complex logistical problem. Deep *air* diving has two additional weaknesses: oxygen toxicity and nitrogen narcosis. Since I have already discussed oxygen toxicity in the chapter on "Accelerated Decompression," I would now like to devote some space to nitrogen narcosis.

Narcotic at any Depth

There is no doubt that an increase in the partial pressure of nitrogen affects the brain, and, consequently, the mind. But in diving, what is sometimes termed narcosis is more correctly expressed as an intensified focus of attention: a concentration on that which is most important at the time, accompanied by a neglect of ostensible minutiae. Think of it as a preoccupation for fundamental values.

Later, when there is time for deep reflection, one consciously becomes aware of those peripheral references. One remembers what went into the buffer zone of one's mind during the span of sensory overload.

This doesn't necessarily imply that a person was narked on the bottom, only that the primary concerns of execution took precedence over secondary observations. The mind can handle only so much incoming data, so it has to choose which data are important enough to require immediate processing, and which can be stored for later retrieval.

I think it likely that divers who handle deep air diving with a minimum effect from narcosis have a natural propensity for selecting which data require prompt consideration, and which can be ignored for the moment. The partial pressure of nitrogen might slow them down, but it doesn't incapacitate them.

*The capacity to manage priorities under pressure
I call the "snapshot facility."*

It's as if the brain were a camera taking a series of quick photos, and the mind an interpreter scanning the pictures as they were taken and prioritizing recall. The mind then separates the images of events into sequential groups of urgency: those that require instantaneous action, those that can afford a postponed response, those that can be placed on provisional delay, those that are delegated to later review, and so on. Information that is nonessential for the nonce is temporarily ignored.

This quick response mechanism applies not only to diving, but to any emergency or high-stress situation in which people must act or react suddenly and decisively. People who possess this natural talent will observe and remember more of their experiences in life than those whose minds are out of focus.

Deep divers work under severe time constraints that are totally independent of the narcotic effect of nitrogen. Thus their powers of observation must be accelerated if they are to achieve their goals under water. If they tend to overlook immaterial events, it is because vital circumstances drew emphasis.

Replacing nitrogen with helium will not overcome such a manifestation of distraction. It only eliminates narcosis.

Helium - It's a Gas

Since both oxygen toxicity and nitrogen narcosis are functions of excessive partial pressure, the obvious solution to both problems is a fractional reduction of the gas responsible for each complaint.

To prevent the onset of oxygen toxicity by decreasing the ppO_2 in air automatically increases the ppN_2, resulting in a greater potential to bring on nitrogen narcosis at depths shallower than those at which it is normally felt. Conversely, lowering the ppN_2 raises the ppO_2, inducing oxygen toxicity at shallower depths. You can't decrease both simultaneously unless you fill in the gap.

Enter helium, the gas of the sun.

Helium's solar-sent properties make it ideal for deep diving: it is inert, non-narcotic, and is not metabolized by the body. On the down side, it is expensive, is absorbed faster in fatty tissues, and makes you sound like a Munchkin when you talk.

The expense can be rationalized, the decompression can be accelerated, and, since talking on scuba is generally kept to a minimum, helium's effect on the vocal chords is hardly an inconvenience. (Of peripheral consideration is the property of a helium mix being not as dense as a nitrogen mix, therefore making it easier to breathe under extreme ambient pressure.)

Myth-Conception

Before I go on, let me correct a common misconception: that the decompression penalties for helium are less than those for nitrogen, and that shorter hang time is the reason that helium is breathed on deep dives.

Quite the contrary, inspired helium imposes not just longer decompression penalties than nitrogen, but significantly longer. Decompression may take so long that one could not possibly carry enough gas to complete a hang.

I could demonstrate this by drawing a comparison between the U.S. Navy Standard Air Decompression Table and the U.S. Navy Helium-Oxygen Scuba Decompression Table, both readily referenced - but that would be dangerous. In addition to the caveats I've already made with respect to Navy Air Tables, the Navy HeO2 Tables are even more impractical for technical diving use.

The Navy blend consists of 32% oxygen and 68% helium. You will remember from my discussion of nitrox that I was quite insistent that nitrox-32 (that is, air in which the fraction of oxygen is enhanced to 32%) must not be breathed below a depth of 130 fsw due to the potential for incurring oxygen toxicity. That the Navy recommends the same percentage of oxygen for breathing at 200 fsw by simply replacing one inert gas with another in no way reduces the risk of oxygen toxicity. That is a function solely of ppO_2, and has nothing to do with the inert gas that makes up the balance of the mix.

Remember the Navy credo that assumes an "acceptable" degree of risk. The Navy blend was designed primarily for combat missions. Navy divers are therefore exposed to the risk of oxygen toxicity with the same nonchalance with which they are exposed to the risk of lead poisoning from enemy gunfire. Part of their job description is working in hostile environments.

Technical divers are under no such obligation. They have control over their lives and their diving conditions, and should exercise caution accordingly.

The only reason that helium is used to replace nitrogen in a breathing mixture is because it is non-narcotic - at least down to depths of 1,000 fsw and beyond the scope of this book. Two helium mixtures are prevalent: trimix and heliox. Let's take them one at a time, in descending order.

Trimix

Trimix is a combination of three elementary gases (oxygen, helium, and nitrogen) plus impurities and trace gases commonly found in the atmosphere. The impurities and trace gases are introduced into the blend any time the partial mix is topped off with air. (See "Expedition Diving" for the specifics of gas blending.)

Trimix comes in a wide variety of fractions, each chosen with respect to the depth and duration of the dive. The partial pressure of oxygen must be kept within certain critical parameters: too high can induce toxicity, too low can lead to shortness of breath, poor judgment, and eventually hypoxia.

The key word in mixed-gas diving is balance.

Mountain climbers, ascending to heights where the air is thin, have long known that the cognitive and decision making processes are adversely affected by the lack of oxygen reaching the brain. They suffer from insufficient ppO2. To overcome this, they sometimes breathe pure oxygen at altitude. The phenomenon is well understood because of extensive testing in the field.

Divers breathing mixed gas have another problem to consider: too high a ppO2. If I mention oxtox too often it's because I'm trying to stress a point. Symptoms of toxicity don't always creep up on you as narcosis does. It is possible to go into a full blown seizure without warning. In that case, unless you're wearing a full-face mask or someone notices your wild convulsions and takes you to the surface, you could drown.

The only good news is that you'll never know what happened. It's like falling asleep and not waking up. But what a waste of the rest of your life.

Depth and Duration

Remember that I wrote that both depth and duration must be considered when choosing the percentage of oxygen for the mix. If we were concerned only about the partial pressure of oxygen, a simple equation would suffice for determining the appropriate percentage of oxygen for the maximum depth of the dive.

But equally as important is the duration of the dive: not just the bottom phase, but the decompression phase as well. When we push the partial pressure of oxygen to its acceptable safe limits, we must take into account the *total* exposure to oxygen. The CNS clock is running during decompression too - sometimes even faster - mak-

ing it possible to have a seizure in only 10 fsw.

This means that if you chose a bottom mix and deco gases in which the oxygen content was based on the threshold extreme of 1.6 ATA of oxygen, you'd probably exceed the allowable exposure of oxygen unless you were making only a bounce dive. But the average running time of a mixed-gas dive is a couple of hours - all while breathing vastly elevated partial pressures of oxygen.

Thus the atmospheres absolute have to be cut back to, say, 1.5 ATA or 1.4 ATA.

Equivalent Narcotic Depth

Another concept to balance the scales of the constituent gases in trimix is equivalent narcotic depth. Since trimix contains nitrogen, it can induce narcosis. The degree to which narcosis affects a person is based on a number of factors: individual tolerance, work load, physical and emotional stress, body temperature, and, of course, the partial pressure of nitrogen.

The primary job of helium in trimix is to reduce the partial pressure of oxygen. Another job is to reduce nitrogen narcosis.

Simply stated, equivalent narcotic depth is the amount of narcosis you would feel from breathing trimix on the bottom compared to the depth at which you would feel the same amount of narcosis from breathing air.

In other words, before prescribing the percentages of nitrogen and helium that will share the inert gas portion of a blend, you decide how "deep" you want to go on nitrogen. That is, what effective depth of nitrogen narcosis you're willing to tolerate. The ratio of nitrogen to helium is based upon that determination.

For example, I made a 230-fsw dive on the *Monitor* breathing trimix consisting of 19% oxygen, 30% helium, the balance nitrogen. In shorthand this is written trimix-19/30. The percentage of nitrogen is inferred, but you can easily calculate that it represented 51% of the gas in the mix. The narcotic effect of that much nitrogen at 230 fsw is the same as that of air at about 140 fsw.

Here's the equation for calculating equivalent narcotic depth:

$$END = \left[\frac{(fN2)}{.79} \times (fswA) \right] - 33\ fsw$$

If this equation looks familiar it's because it's the same one used to calculate the equivalent air depth of a nitrox blend. In this equation (fN2) is the (fraction of nitrogen) in the mix, and (fswA) is (feet of sea water absolute), or, if you prefer, (D + 33) in which D is the depth.

Let's use trimix-19/30 as an example. To get the fraction of nitrogen we add the fractions of oxygen (.19) and helium (.30). Subtract .49 from 1.0. Thus fN2 is .51. Divide that by .79 and we get .6455. Multiply .6455 by 263 (the sum of 230 fsw actual depth and 33 fsw absolute), and the product is 169.77. From that subtract 33. The END is 136.77 fsw.

All of this means that when I reached the sand surrounding the *Monitor* I felt the same as I would feel at 140 fsw on air. And that's the beauty of mixed gas diving.

Check it Out

Other factors that went into determining the optimum blend to breathe on the *Monitor* were the CNS clock and the different decompression requirements for both inert gases. This profile was generated by Bill Hamilton's decompression program called DCAP. More on this later.

Let me give you another example.

It may shock you to learn that my first trimix dive was made to 250 fsw. The *Wilkes-Barre* was my testing ground, with Billy Deans, owner of Key West Divers, as my adviser. This may sound pretty deep for a check-out dive, but a shallow dive on trimix doesn't demonstrate the virtues of breathing helium.

In order to contrast the effects of narcosis I first made a control dive on air. I felt slightly out of focus, and as usual on a deep dive I perceived the stern apprehension that was exaggerated by my knowledge of the press of time, the limited air supply, and the long decompression penalty. This is serious diving and is not to be taken lightly.

The next day I made a comparison dive on trimix. The difference, to say the least, was startling. While my performance on air had not been poor (due to my experience and, admittedly, a tol-

erance to narcosis derived largely from heredity), my powers of observation were greatly enhanced.

A case in point. A fish trap snagged on bottom wreckage was still catching fish. I wanted to open the door in order to release the captured denizens ere they died. On air I couldn't figure out the latching mechanism. On trimix I was able to spot the problem. Marine encrustation cleverly disguised the hinges and wooden slide. Once I could focus my concentration without the affect of narcosis, I easily twisted the knob and let the trapped fish escape.

Note that on air I didn't have trouble reading gauges, removing or replacing my knife, adjusting buoyancy, or executing any other action or maneuver. But when challenged by an unfamiliar contrivance I was stymied.

I felt as comfortable and as clear-headed on that 250-fsw dive as I was on a shallow reef. And there's the rub: normally I can *feel* the depth. I can taste it in the air. On trimix I felt no sensation of depth, felt none of the apprehension I'm used to relying upon to keep me from being too bold.

This false sense of security was overcome only by intellectualizing my vulnerability: not only was my gas supply limited, but return to the anchor line was paramount because the decompression schedule penciled on my slate called for switches to deco gases that were either hanging in bottles on the anchor line or were fed through hoses from a storage cylinder on the boat.

Only by looking at my depth gauge could I comprehend what I couldn't feel. It had never occurred to me that clear-headedness could be a drawback.

Liquid Courage

A wreck like the *Wilkes-Barre* is an ideal location for trimix check-outs. So is an open spring like Hal Watts' Forty Fathom Grotto. Both sites offer relative peace of mind for a mixed-gas trainee who has his share of trepidation to deal with. Both sites also have invisible handicaps.

The *Wilkes-Barre* offers phenomenal visibility. Forty Fathom Grotto presents a confined, controlled environment. The vistas and the womb are comforts that

Forty Fathom Grotto.

a trainee needs. But these accommodating qualities and the abeyance of narcosis invite "liquid courage." They mask to some extent what breathing helium is all about.

Don't be fooled by the effortless travail, when narcosis no longer opaques your thoughts or clouds your normal judgment. When there's no one there to hold your hand, when you leave the pond for the greater world beyond, when you descend to a wreck where the hazards are real, when you explore a cave that is long and sinuous, that is where you learn what mixed-gas diving is truly about.

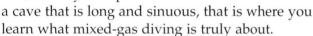
The stern of the USS *Monitor.*

No matter what you're breathing, deep is deep. The decompression penalties are severe, and the slightest error or miscalculation can get you into serious trouble. Do not underestimate the potential perils and pitfalls of extremely deep diving.

The Deep Range

For most people, the range for diving on air spans from the surface down to 160 fsw to 200 fsw, the depth at which narcosis begins to exert a negative influence. Trimix comes in at a depth where oxygen toxicity rears its ugly head. This is not a mental barrier but a true physical limitation. Heliox, discussed in the next section, is for *deep deep* diving.

What used to be considered deep is now the stepping off point for truly deep.

Mixed-gas diving (or helium diving, if you prefer) has much to offer. It enables you to go places you couldn't reach before, with a head that is clearer than your conscience. Lest helium diving sound too attractive, however, understand that it requires a sacrifice of time and money that not everyone is willing to make. In order to extend your underwater range you need training, equipment, experience, dedication, and most of all, initiative. Helium diving is incredibly labor intensive.

Let's exercise some mental gymnastics and explore the complicated logistics that are involved in pulling off a successful dive, and a successful dive trip.

Pick a Mix

When I organized the 1990 USS *Monitor* Photographic Expedition, I considered five alternative gas mixes recommended by Billy Deans: trimix-17/50, trimix-17/36, trimix-17/25, trimix-17/17, and the one we ultimately breathed, trimix-19/30. Each blend offered a different equivalent narcotic depth, ran to different percentages of the CNS clock, and required its own decompression schedule.

Factors that I weighed prior to choosing the actual mix included not just END, CNS percentage, and total in-water decompression penalty, but the planned duration of each dive, the possible number of consecutive dives, the number of days the expedition was expected to last, and the availability of space on the boat for tanks (for bottom mix and deco gas).

The EAD of trimix-17/17 was 187 fsw, which I thought was too deep for the mental clarity I needed for what I set myself to accomplish. The decompression schedule for trimix-17/50 called for three gas switches (nitrox-51 at 70 fsw, nitrox-72 at 40 fsw, and oxygen at 20 fsw); these gases optimized the efficiency of decompression but added to the logistical complexity. These gas blends were the extremes.

Since I intended to carry two cameras on every dive, I went for simplicity and chose a middle ground. Trimix-19/30 yielded an END of 137 fsw, required only two deco gases (one of which was surface supplied oxygen), and compared favorably in decompression profile with accelerated air tables. I carried only one supplementary tank.

On the other hand, when I explored Dipolder and Eagle's Nest with veteran cave diver Steve Berman, he suggested a "light" mix and two deco gases. We breathed trimix-17/17 with its relatively deep END because we intended only to sight-

AIR AND TRIMIX COMPARISONS
240 fsw for 25 minutes

GAS	DECO	END	CNS
air	55 min	240 fsw	275%
17/27/56	75 min	161 fsw	50%
17/36/47	76 min	129 fsw	50%
17/45/38	77 min	98 fsw	51%

The table above is intended for comparison purposes only and should not be used in actual practice. All profiles require an initial stop at 90 fsw. Deco times are based on deep stops breathing nitrox-36, then switching to nitrox-70 at 40 fsw. In the GAS column, the sequential values represent the fractions of oxygen/helium/nitrogen. All information was provided by Ted Green.

In this example the differences in deco time among trimix blends are insignificant, while the reduction in equivalent narcotic depth is great. Furthermore, the CNS clock for air runs to an outrageous 275%. Once the decision is made to breathe trimix, the cost of additional helium for blends with shallower END's is trivial.

Note that a 20-minute bottom time would reduce the deco time markedly.

TRIMIX END COMPARISONS

DEPTH	END 100	END 130	END 160
150 fsw	25/18/57	25/5/70	air
180 fsw	21/30/49	21/18/61	21/7/22
210 fsw	19/38/43	19/28/53	19/18/63
240 fsw	17/45/38	17/36/47	17/27/66
270 fsw	15/50/35	15/43/42	15/35/50
300 fsw	14/55/31	14/48/38	14/40/46

The table above is intended for comparison purposes only and should not be used in actual practice. Trimix blends are written sequentially as oxygen/helium/nitrogen, in order to show the ratio of helium to nitrogen in the blend. Thus 25/18/57 is trimix-25/18. The calculations are based upon a partial pressure of oxygen of 1.4 ATA. Fractions were rounded up for ease in reading. All information was provided by Ted Green.

The difference in decompression times between equivalent narcotic depths for any given actual depth is in the order of 5 to 10 minutes per hour of total decompression. Such a slight fractional increase in run time advocates choosing an END that is shallower rather than deeper. A clearer head can be bought at the expense of a few minutes of hang time.

see: there was no responsibility to perform at depth. At 275 ffw the small fraction of helium offered a shorter total ascent time than a blend with a shallower END.

Nor was carrying two side-slungs disadvantageous. We staged the deco tanks on the way in so we didn't have to swim with them during the exploration phase of dive. We picked up the tanks on the way out.

Thus you can see that a variety of factors are considered before deciding upon which mix to breathe and how many gas switches to make. There is no right way and no wrong way, but you must plan your mixes for the prevailing conditions of the dive.

Heliox - for Really Deep

Heliox is a blend of two gaseous elements, helium and oxygen, and is a purer blend than trimix. Not only does heliox contain no nitrogen, but it doesn't contain the impurities and trace gases that are commonly found in air. This is because heliox is made by mixing oxygen with helium, each in its pure form. No air is introduced into the blend. (See "Expedition Diving" for the specifics of gas blending.)

While trimix is breathed most commonly at moderate or mid-range depths, say, less than about 300 fsw, heliox is breathed on dives that go beyond. Far beyond.

The advantage that heliox has over trimix is the total absence of nitrogen in the mix. This means that narcosis is reduced to zero. By applying that stupid martini analogy that used to be given in basic certification courses with respect to nitrogen narcosis (every 50 fsw of depth is like drinking another martini) breathing heliox is equivalent to drinking a glass of Coca-Cola.

The complete lack of narcosis doesn't do away with anxiety, stress, or other abnormally heightened emotions that are normally ascribed to nitrogen narcosis. The pressure of time and the distance to the surface induce a certain amount of tension that is normal to feel.

Your head is clear, your judgment unimpaired. Nevertheless, you can't make any better decisions on the bottom than your training and experience allow - or that you're inherently capable of making on the surface. Helium doesn't administer intelligence and common sense, it just doesn't take any away.

Gas Management and SAC Rate

Divers who are used to breathing air at air depths know how long they can remain on the bottom and still have enough air for decompression. Large capacity tanks fulfill the need for those who have high consumption rates. People can pretty much go with the flow.

But diving on helium demands a bit more planning and foresight. The abyssal plane is no place to run out of gas. As everyone knows from basic scuba, the deeper you go the more gas you breathe. Extrapolate this adage to 300 fsw or more and it takes on a whole new meaning. Strict attention to gas management is not a negotiable concept.

So how can you anticipate the quantity of gas you'll need? Going with the flow isn't good enough. You must apply scientific methods, then toss in some fudge factors for an added margin of safety.

A good way to begin is by calculating your surface air consumption rate (SAC rate). On a routine dive, while decompressing at the 10-fsw stop (scuba divers rarely float on the surface), watch your tank pressure gauge and note the reduction in psi during a fixed amount of time. From the table of tank capacities in Chapter 1 obtain the cubic feet of air that the total change in psi represents. Divide that amount by the number of minutes, and you have an approximation of the quantity of air you breathed.

For all practical purposes this is your SAC rate *at rest*. In fact, the concept of surface air consumption rate assumes that you're at rest. It doesn't account for how cold you felt, how hard you worked, or how nervous you were. If you multiply your SAC rate by atmospheres absolute, you can figure about how much air you'll breathe at any particular depth. In equation form it looks like this:

$$\text{air consumption} = (\text{SAC rate}) \times \frac{D + 33}{33}$$

D is the actual depth. If your SAC rate is .75, at 99 fsw you would consume 4 cubic feet of gas per minute. At 165 fsw, 6 cfg/minute. At 231 fsw, 8 cfg/min. At 330 fsw, 11 cfg/min.

This calculation doesn't account for time of descent or ascent. You can get an approximate travel consumption by averaging the depth, but most people simply count travel time as bottom time and round off the difference to conservatism.

It's also possible to estimate your *working* consumption at depth. Calculate how much air you consume during the bottom phase of the dive. Follow the procedure already described: note the tank pressure upon arrival and retreat and figure it out from there. Write the numbers on your slate if you have trouble remembering.

Repeat the process for a number of dives for various depths and conditions. The average will be a measure of your working consumption at depth. Use the equation above but solve it for your working SAC rate.

Now you have two consumption rates: one resting and one working.

If your helium dive calls for invoking the thirds rule, plan accordingly.

Helium Management

Air consumption and helium consumption are different. Despite what anyone says, psychological stresses exist at depth. Being so far away from the surface is scary. I have no problem admitting it.

There are also physiological reasons for consuming helium faster than air. No one knows why for sure, but according to Bill Hamilton, "This is probably due to the body's tendency to increase ventilation rather than allow buildup of CO_2, and this is easier to do when the gas is less dense." You don't have to understand it, just accept it. "Some divers have reported an increase in gas consumption of 20 to 25% when using trimix as compared to air."

Although this evidence is anecdotal, take note that the anecdotes originated from experienced helium divers, not from trainees after their first deep descent.

Another factor to keep in mind is that you can't breathe all the gas in your tanks. It won't all come out. During my return from a 420-fsw dive I had my eyes glued to my gauges. The tank pressure gauge showed 500 psi as I approached the 300-fsw mark. I was congratulating myself that I was going to make it to my gas switch depth on my doubles when suddenly my regulator stopped delivering. From the first notice of trouble till I got no more gas was the breadth of, well, a breath. Then no more gas.

Frantically I scrambled for my auxiliary tank

of bottom mix, and shoved the regulator into my mouth. I breathed a deep sigh of relief. What was the problem? The first stage couldn't overcome the ambient pressure. This is partially due to the regulator model, partially due to the extremity of the depth.

One-seventh of the gas in my double 120's was unavailable to me at the depth where I needed it. This is a fact to consider when predicting the quantity of gas you will need.

A Memo from the Management

The importance of gas management assumes new proportion when the factors noted above are given heed. Consideration must be given to consumption rate at rest, consumption rate under load, emotional duress, accelerated breathing pattern, gas held in reserve, and unreachable gas in the low pressure zone of your tanks.

Although personal computer software programs generally calculate gas consumption, there's no equation that can factor in so many variables and yield a meaningful solution. Only personal experience will provide the answer.

German Battleship Ostfriesland

Let's look at the logistics of a truly deep dive. The German battleship *Ostfriesland* was sunk by General Billy Mitchell's army planes in a bombing test off the Virginia Capes in 1921. The wreck lies at a depth of 380 fsw, far offshore, in an area where the bottom conditions were completely unknown in 1990, when the first expedition to the site was conducted, by Ken Clayton, Pete Manchee, and me.

The challenge of descending into the realm of unexplored depths and in an uncontrolled environment compelled an ultra conservative approach. We conferred with Bill Hamilton and asked him to provide a decompression profile suited to our needs. He recommended a breathing gas of heliox-12: that is, 12% oxygen, the balance helium. The degree of narcosis induced by breathing such a mix on the bottom is about the same as that of breathing air in your living room.

There were two shortcomings with this blend: the length of the decompression requirement produced by the high percentage of helium, and insufficient partial pressure of oxygen near the surface.

Decompression penalty is a function of depth, duration, and inert gas absorption: quanta that are absolute and therefore immutable. Decompression on heliox is exceedingly long because of helium's high absorption rate. If you want to dive deep, you have to hang your laundry out to dry.

The only way to shorten the decompression profile was to incorporate multiple deco gases. There are an infinite number of variations to this scheme. The trick is to find a compromise between the number of deco gases and the increasing partial pressure of oxygen, and balancing the act by controlling the CNS clock.

You will remember from the "Introduction" that storms can kick up quickly off the Virginia coast. We didn't want to extend our in-water decompression any longer than necessary. But there's a limit to the number of deco bottles a person can handle, particularly in the open ocean.

Dovetailed with these decompression considerations was the low percentage of oxygen in the blend. The percentage was chosen to yield the appropriate partial pressure of oxygen on the bottom relative to toxicity and the CNS clock. But on the surface and during initial descent, where we were likely to encounter waves and strong current on the way to the anchor line, 12% oxygen might not satisfy our respiratory requirements.

Hamilton treated both dilemmas by recommending a travel mix. Thus we breathed heliox-16 on the surface and down to 320 fsw, at which point we switched to our bottom mix. Upon ascent we switched back to the travel mix at 150 fsw, so that heliox-16 was our first deco gas. At 110 fsw we switched to nitrox-36. At 20 fsw we went on pure oxygen that was surface supplied.

I went into the water wearing double 100's on my back and two 80-cfg singles slung to my side. One single was filled with travel mix, the other with bottom mix. I had no intention of running

out of gas on the bottom. I also carried a video camera. So where did I carry the nitrox for deco?

The nitrox-36 was staged on the anchor line at 110 fsw. Furthermore, we were each assigned a support diver who descended to 110 fsw at the predicted time of our arrival, to help unclip the side-slungs and take them to the boat.

In retrospect, we had more gas than we needed, but we didn't know that at the time and were unwilling to take the chance of running short. It is ironic to note that our first decompression stop was 160 fsw: deeper than most divers have ever gone.

Nowadays, we've done away with the two different helium mixes and breathe nitrox (a deco gas) as a travel mix.

Soporific Nitrogen

Some divers have reported feeling a sense of unease when breathing heliox, but claim to feel more relaxed when breathing trimix at the same depth. Psychological factors notwithstanding - everyone should feel uneasy at 400 fsw - it's demonstrably true that some partial pressure of nitrogen buffers aggravating attacks of anxiety.

We tend to think of nitrogen in terms of debilitating narcosis. But we should also comprehend that a narcotic whose dosage is reduced can be a beneficial sedative. I suspect that those who complain of these feelings of apprehension are perceiving blatant reality through the eyes of an unanesthetized brain.

Heliox versus Trimix

The trend in helium diving has taken a turn away from heliox since those early days on the *Ostfriesland*. As noted in the previous section, trimix, with a small percentage of nitrogen acting as a mild tranquilizer, tempers that uncertain feeling. Let's hope the Food and Drug Administration doesn't find out, or the prohibition on controlled substances might be expanded to include breathing nitrogen at depth.

Heliox still has its uses, not only in extremely deep depths (say, 500 fsw and beyond) but for shallower dives in which intense concentration on mission goals requires the complete absence of narcosis.

Differences in decompression profiles between the two helium mixes do not favor one over the other for the majority of technical diving applications. There is a common misconception that breathing heliox entails longer decompression penalties than breathing trimix. This is true and false. What makes the difference is duration.

For long dives (more than an hour) breathing heliox and appropriate deco gases yields shorter decompression penalties than breathing trimix and appropriate deco gases. Bill Stone and Bill Hamilton offer a partial explanation: "While nitrogen dissolves at a slower rate than helium, it also out-gasses at a slower rate; helium, while reaching a higher percentage of saturation, also off-gasses at a faster rate. When nitrox is used as a decompression mix it will actually accelerate the off-gassing of helium, through isobaric counter-diffusion. There is no corresponding acceleration of the off-gassing of the nitrogen portion of the mix."

For short dives (twenty minutes or less) breathing trimix and appropriate deco gases yields decompression penalties nearly identical with those of heliox. In between twenty minutes and an hour is a gray zone in which penalties are comparable, depending upon how much helium is in the three-gas mix. Trimix with about 50% helium or more is equivalent in decompression penalty to heliox.

Except for a few highly organized underwater cave expeditions, most technical dives fall into the short range category.

The primary reason that trimix triumphs as the mix of choice is one of simple economics: nitrogen is less expensive than helium. Granted that there's a trade off between helium and equivalent narcotic depth, the fact of the matter remains that trimix blends are more cost effective than straight heliox.

The Argon Offensive

Since we wore drysuits on the *Ostfriesland* dive we used argon as our inflation gas. I carried mine in a pony bottle that was hose-clamped to the back of my doubles.

Contrary to popular belief, the loss of heat caused by passing helium through the lungs and respiratory tract is not as great as it is with air. *But*, being immersed in a helium atmosphere increases heat loss considerably.

Pumping helium into a drysuit makes for a cold dive, whereas the density of argon conducts heat away from the body at bureaucratic speed.

Argon also makes a person profoundly warmer on air dives in cold water, such as on winter ice dives or in cold, fresh water lakes and northern caves. It can make the difference between comfort and discomfort or between shivering and shaking violently.

The virtues of argon as a warmth provider should not be overlooked.

How much argon do you need for a dive? It depends upon such variables as the depth of the dive, the size of your drysuit, and your extravagance with the gas. It does *not* depend upon the temperature.

• Even though you rely on a BMF to provide the majority of your flotation, a certain amount of inflation gas is necessary to overcome suit squeeze. That amount is dictated by ambient pressure - back to basics with good ole Boyle's law: the deeper you go, the more gas you need.

• Large people wear large drysuits. Sure there's a body inside taking up some of the extra room, but there's a lot more dead space between the skin and the material.

• A loose fitting drysuit requires more gas to inflate initially, but not after equilibrium is reached.

• Gas that is burped around a drysuit's seals must be replaced. Make sure that your seals are snug and are not stretched out of shape.

• If you flush your drysuit frequently during pre-dive warm-up, you waste some of the gas that you could be saving for depth. Obtain argon for pre-dive flushing from a different bottle than the one you plan to take with you into the water.

• Overinflation to the point that requires venting wastes gas.

• Venting too much gas during ascent requires replacement of vented gas. Practice control.

I repeat, how much argon do you need? I will answer by way of example. On a dive on the *Norness* at 280 fsw I carried argon in a newly obtained 6-cfg aluminum bottle. I ran out of inflation gas just as I touched the bottom. I wore a BMF for buoyancy control so my concern was the inability to replace vented argon. I continued the dive but, fearing cold and temporary suit squeeze, I vented judiciously during ascent.

I determined that the bottle I carried was not

adequate for my needs. Its fill pressure was 3,000 psi. Argon is dispensed from storage cylinders that are pressurized to 2,400 psi. The argon in my bottle was cascaded from a storage cylinder that was less than full. When I started the dive, the gauge pressure on my argon bottle was only 1,800 psi. Subsequent cascaded fills were even lower.

If my bottle had been full it would have sufficed for that depth (but not necessarily for a deeper depth). It was impossible to fill the bottle to capacity by the cascade method. In order to carry the quantity of argon I needed for such I dive, I had to have either booster capability, a larger bottle, or a bottle with a lower fill pressure.

I eventually opted for a 13-cfg low-pressure steel bottle. The small aluminum bottle I reserve for shallow, cold water dives.

The *Norness* trip lasted for three days at sea. For later dives I topped off my argon with air from the boat's compressor. This provided enough inflation gas for a dive, while only partially reducing the insulating qualities of the argon.

(The cascade system is explained in the chapter on "Expedition Diving.")

Warm Water Inflation Gas

Argon is great if the water is cold, but how about if it's warm? Wear a wetsuit is the answer that immediately pops to mind. Nor do I have a problem with that, even for long exposures. But most technical divers prefer to dive dry for reasons other than warmth, such as comfort, ease in dressing, back-up flotation, and because a drysuit is an integral part of their system.

There's nothing wrong with wearing a wetsuit on deep or extended range dives. I make no such implication. But the fact remains that most technical divers have gotten used to diving dry. To switch outer wear from one dive to the next is like changing horses in midstream. Once you get used to drysuit diving and rig your gear accordingly, deviating from routine is more trouble than it's worth.

If you're going to dive dry in comparatively warm water, ask yourself first if argon is necessary for suit inflation. If not, consider putting air in your argon bottle. Some divers dispense with the inflation gas bottle altogether by connecting their inflator hose to their deco gas tank. There's nothing wrong with pumping your suit full of

nitrox from the sling bottle at your side - unless, of course, you plan to stage the bottle. Your BMF can be inflated in similar fashion.

Take the battleship *New Jersey*, for example. It lies in tepid waters south of the Diamond Shoals of North Carolina, at a depth of 330 fsw. When I dived the wreck, the temperature on the bottom was 66° F and at decompression depths it was 78° F. This is clearly a wetsuit dive - perhaps a shorty would do. But I like diving dry.

So I wore a pair of thin longjohns and filled my drysuit with bottom gas: trimix-13/50. The helium provided a cooling effect that kept me comfy throughout the dive. Nor did I need an auxiliary tank. I pumped helium from my doubles the same as I would if I were diving on air.

It has been alleged that helium can react with the skin to produce a rash or itchy blotches, a condition known as skin bends. It didn't happen to me. Nor does it happen to commercial saturation divers who work out of bells and live in chambers for days or weeks at a time.

Helium Tables

Helium decompression tables exist. But they are either too dangerous to use (such as the Navy tables mentioned above) or closely guarded corporate secrets. Oil rig and commercial diving companies have their own confidential tables, which were developed through years of private research, the careful collection of data in the field, and stern statistical analyses. Such information is strictly proprietary.

Furthermore, those tables fall into the same category as Navy air tables: the design criteria demand the employment of an on-site recompression chamber, an expense that most technical divers can hardly afford.

Besides, it is impractical to publish an all-encompassing set of tables for every mix variation. A fractional change in any constituent gas in trimix and heliox necessitates a unique decompression profile. The decompression requirement for a specific mix, depth, and duration can be derived only from a computer program that is based on a reliable and conservative algorithm that takes into account the elimination rates of the inert gases, the partial pressure of oxygen, and the running of the CNS clock.

You also need faith in the system because each dive is somewhat experimental. That is, a decompression profile generated for a particular combination of mix, depth, and duration might never have been tested by a human subject in the water. Each profile is an interpolation made by a computational model.

Any single decompression profile is called a schedule, not a table.

These schedules are not printed in table format, but are generated on a one-time basis depending upon the parameters of the proposed dive.

In the Beginning . . .

Technical divers today can thank Bill Hamilton for writing the first decompression program made available to the public. He called it DCAP (pronounced dee-cap), which stands for "Decompression Computation and Analysis Program." DCAP was used extensively by cave divers in the 1980's. Wreck divers began using it in 1990.

DCAP's most attractive feature was its user friendly operation. The only variable that a diver had to supply was the predicted maximum depth. Hamilton provided all the other data: bottom mix, travel mix, deco gases, switch depths, ppO₂ range, and of course the decompression schedule. With cover material, explanatory notes, printed schedules, references, and appendices, the package was over forty pages long.

The cost of this package was $400 for each designated depth. That's how much we paid to have a schedule generated for the _Ostfriesland_ dive. Part of the reason for the high cost was the time it took Hamilton's computer to calculate and print the required data once he had input the parameters.

The user needed no special training or experience to obtain the full particulars of the dive - only Hamilton's address or phone number.

Laptop Decompression Software

On the market today are at least a dozen decompression software programs that can run on home computers and battery powered laptops. These programs can spit out in seconds what took DCAP hours. Furthermore, these programs are so user friendly that anyone who can walk and chew gum can learn to run his own decompression

schedules without resorting to professional programmers.

This is not to disparage DCAP in any way. Prototypes are often considered crude by later-day standards, after second generation models have been developed from original designs and concepts. No one who has seen the _Spirit of St. Louis_ in the Smithsonian Institution would consider flying such a crate across the Atlantic Ocean now that commercial jetliners offer luxury flights much faster. Yet Lindbergh's old-fashioned single-engine propeller plane was once considered aviation's most modern machine.

It is meaningful to note that during the initial proliferation of home deco programs, the validity of each was justified by how closely its profiles compared to those generated by DCAP. That almost says it all. In addition, DCAP's computation speed was limited by the central processing units that were available at the time. Modern personal computers are driven by microprocessors that operate substantially faster than mainframe computers of a decade ago.

Today DCAP can sprint with the best of them. Nor is DCAP outmoded in any other way. In fact, the program has been licensed to drive the Cis-Lunar rebreather.

One oddity that you might find disturbing is the dissimilarity of generated schedules that a comparison among programs reveals. Each program is unique in that it was written by a different programmer, is based upon a distinct algorithm, and incorporates its own brand of chocolate fudge factors. Which one is absolutely correct? None, but they're acceptably close.

Running a Home Deco Depot

Like any other software package, a home deco program seems complicated at first, even after reading the instruction manual. But it gets easier to use after you play with it for a while. And you don't even need to know how to touch-type because the input data are entered by single key strokes. Hunt-and-peck will suffice.

Mac users can start screaming now. The deco programs currently on the market operate only on DOS, usually with a Windows interface. Promises have been made, and broken . . .

Individual software packages vary the same as any other product. The generic program is defined with default options for such parameters as partial pressures of oxygen, the CNS clock, oxygen toxicity units, rates of ascent and descent, equivalent narcotic depth, and so on.

Each default option can be modified. The user can raise or lower the ppO₂, OTU's, END, CNS clock, and any other initialism you can think of. You can also vary the stop depths and the rates of ascent and descent. Next, the user plugs in gas blends for bottom mix and deco gas (which are suggested by the program based upon the ppO₂ selected), and a unique dive profile appears on the screen.

You can customize the schedule by inserting conservatism factors in the form of increased percentages in the decompression penalty (5%, 10%, and so on), SAC rate and anticipated workload, thermal characteristics (is the water warm or cold, are you diving wet or dry?), your physical condition, even your age, weight, height, and sex (male or female only, not orientation or frequency).

Be careful. It's possible to define a schedule so conservative that you'll never get out of the water. You'll have to live forever at 10 fsw.

SAC rates are important to enter because the program can then predict the quantity of each gas you'll breathe on the bottom and for deco. Then you can plan on what size tanks to carry.

These programs can be used for _any_ blend of gas: trimix, heliox, nitrox, even air. They can generate all the decompression schedules described in the previous two chapters. For an air dive with accelerated decompression, define the bottom mix in percentages (21% oxygen, 79% nitrox), with nitrox and/or oxygen as deco gas. After all, in mixed-gas shorthand air can be written as bimix-21/79.

And what's the cost of such a super deco software package that does everything but wash the windows? Less than what we paid for a single depth profile to dive the _Ostfriesland_.

Load the software onto your laptop, take it with you on the boat or to the cave site, and you can cut a last-minute decompression schedule any time you choose. If your gas percentages are off, if the depth turns out to be deeper or shallower because of a change in destination, if you make a repetitive dive, you've got the essential tool to make appropriate and accurate adjustments.

Get yours today. The possibilities are limitless.

Run Time

Back when I used Navy Tables I was forever losing track of my stop times. I'd reach my first decompression stop, glance at my watch, then, as the minutes ticked by and my mind wandered, I'd forget exactly when I arrived. Was it 43 minutes or 46 minutes? Was I supposed to go to the next stop at 56 minutes or 53 minutes. From what point was I supposed to count the time for the stop? And this when I made only three-stage hangs.

I'd be completely befuddled on a sixteen-stage hang, like that required for the _Ostfriesland_. But most deco programs take all the guesswork out of stop times by adding a run time format to the schedule. This was designed with people like me in mind. Run time is like a countdown in reverse.

Run time is vernacular for running time. It has the same meaning as elapsed time or flight time or mission time. Run time bears no relation to Olympic marathons, in which the run time is a fixed and final number of minutes and seconds, but is instead an ongoing measurement of the passage of time since the initiation of descent.

For example, the accelerated profile for a 20-minute air dive to 200 fsw might give you figures like these: 200 - 20, 140 - 22, 80 - 24, 60 - 26, 50 - 29, and end with 15 - 55. The first number of each pair is the depth, the second number of each pair is the number of minutes into the dive that you're supposed to leave that depth.

If your bottom timer started counting when it hit the water, you'd make your descent and exploration until it registered 20 minutes, at which time you would begin ascending at 30-fsw per minute. The 140 - 22 is not an actual stop but an intermediate waypoint to help you gauge your ascent rate. If you reach 140 fsw too soon (before 22 minutes) it means that you ascended too fast. Wait there until your timer clicks over to 22, then continue upward. If you haven't reached 140 fsw by 22 minutes, hurry up and get with the program.

Likewise, 80 - 24 is an intermediate waypoint. The first true stop is 60 - 26. Rising from 80 fsw at a normal rate of ascent will put you at 60 fsw before 26 minutes. So you hang there until your timer registers 26 minutes. Rise to 50 fsw and leave at 29 minutes. I've left out the next three stops, but you get the idea. You leave 15 fsw at 55 minutes into the dive and ascend to the surface. The dive is over.

The run time method effectively turns your timer into a placement device. By noting the time, you know how deep you're supposed to be. In the example above, if you saw that you were 28 minutes into the dive, a glance at the schedule would tell you that you should have left the 60-fsw stop and be waiting at the 50-fsw stop. You don't have to trust anything to memory.

As an alternative, some divers prefer to carry a stop watch and begin the run time when they leave the bottom. This method is useful if you have to spend any time floating on or just below the surface, in which case an automatic bottom timer might turn itself on and initiate run time before you begin your actual descent.

Some reasons for delaying descent might be swimming from the bank across a broad spring to a vertical cave entrance, placing stage bottles in shallow water, assembling a dive team, and doing in-water equipment checks.

Print, Slate, and Laminate

Mixed-gas decompression schedules are so long and complicated that the average person cannot commit them to memory, especially when a choice of bottom times is provided. After your computer displays a schedule on the screen, you have to transfer the information to a medium that you can take with you on the dive.

The obvious and most often used device for this purpose is the old-fashioned plastic slate, especially if you're cutting a schedule on site. Pencil in rows and columns for the depths and times, then fill in the blanks by transcribing the figures displayed on the screen. A curved slate can be worn on the wrist, a flat slate can be clipped to any convenient D-ring as long as it stays *within visual range*.

The last three words are italicized for a reason. Some people tuck the slate into a pouch or pocket. That is where your back-up slate should go. Others clip the slate out of sight, then have to unclip it and hold it in order to read it. My buddy once dropped her slate as soon as she unclipped it. Since she wasn't carrying a spare, she stayed very close to me for the entire decompression.

You don't have to learn this lesson for yourself; it has already been learned. If you can't read your slate without unclipping it, add a length of line between the slate and the snap hook. And always carry a back-up.

Designate a column for writing in the bottom mix and deco gas. This will remind you what you're supposed to be breathing at any particular depth. Thicken the lines where gas switches should be made or draw the lines in color. Or lightly tint the rows for deco gas in pastels. Or be creative in your application of mnemonics. Switching gases at the proper time is imperative.

Some divers write their deco schedules on duct tape with indelible ink, then apply the tape to the upper surface of their fins. They only have to look down in order to read the schedule. Caution: use only fresh tape that is sure to stick, and don't rely on this method in poor visibility.

If you're running schedules at home in anticipation of upcoming dives, here's a convenient alternative. Print the schedules you need, waterproof the paper by laminating it, then insert a grommet so you can add a lanyard or snap hook.

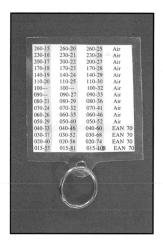

260-15	260-20	260-25	Air
230-16	230-21	230-26	Air
200-17	200-22	200-27	Air
170-18	170-23	170-28	Air
140-19	140-24	140-29	Air
110-20	110-25	110-30	Air
100---	100---	100-32	Air
090---	090-27	090-33	Air
080-21	080-29	080-36	Air
070-24	070-32	070-41	Air
060-26	060-35	060-46	Air
050-29	050-40	050-52	Air
040-33	040-46	040-60	EAN 70
030-37	030-52	030-68	EAN 70
020-40	020-56	020-74	EAN 70
015-53	015-81	015-108	EAN 70

Mark and tint the printed schedule so it appears the way you want it. Trim the edges of the paper to its smallest possible size. For the sake of efficiency, place two schedules back to back before laminating them. Back-to-back pages may be schedules for different depths, different times for the same depth, or one depth with the two sides showing repetitive and non-repetitive profiles. Whatever works for you.

Do not punch the grommet hole through the paper. This compromises the watertight integrity of the lamination and will allow water to seep under the grommet and spread throughout the paper, causing discoloration, fading, or running ink. Also, do not trim the paper *after* it has been laminated; this lets water flow in from the edges. Trim the paper first, then laminate it. You can trim the lamination as long as you don't cut it too close to the paper.

Most copy stores have laminating facilities. If you have a large number of schedules to laminate or would rather work in the comfort of your home, purchase your own laminator or share the cost with your buddies. Standard laminating pouches are three to five millimeters in thickness. This is too flexible for our purposes: the paper bends and folds so it's difficult to read. Buy eight to ten mil pouches, or run the paper through the laminator two or three times until it feels sturdy.

Leave a tail in the lamination for the grommet. Any good hardware store will have a grommet tool and kit. Grommets should be large enough for snap hooks. Punch the hole, pound in the grommet, and you're ready to go.

If the printout is too small to read under low light conditions or without straining hyperopic eyes, try this gambit. Copy the schedule onto your hard drive, paste it into a word processing document, then enlarge the type size or choose a different font.

Another trick is to wrap the decompression profile around the battery pack inside a clear plastic light housing.

Procuring Helium

Mixed gas isn't free as the air. In fact it's quite expensive. The cost is dependent upon two major factors: the price of helium and the labor involved in mixing it to produce a particular blend.

There's not much you can do about the price of helium. It isn't sold in bargain basements and it doesn't go on sale. There may be discounts for quantity purchase, but not in the portions that technical divers use, unless you're a heavy breather or do an awful lot of diving. At least you don't have to pay through the nose; you pay only through the mouthpiece with every deep breath you take.

At the time we made the first *Ostfriesland* dive there was no such thing as a technical dive shop. We purchased heliox from a commercial gas supplier that mixed it to our specifications. If this

is the route you decide to take, read the chapter on "Expedition Diving" and look up gas suppliers in the telephone directory.

But keep in mind that the gas is delivered in storage cylinders. You then have to transfer the gas to your scuba tanks. This requires some capital investment in hoses, adapters, and, if you really want to get sophisticated, a compressor and a booster pump.

If you can't be bothered with the mechanics of gas blending, today there are alternatives in the form of specialty dive shops. There may not be one in your neighborhood, but if one can be found within a hundred mile radius it's worth your while to patronize it and pay the extra money for mileage and mixing.

It takes longer to put gas in a scuba tank than it takes to put gas in a car's fuel tank, or to fill scuba tanks with air. The operator must be in attendance throughout the whole operation, checking partial pressures and opening and shutting valves. The job cannot be rushed. A fast air fill results in a short fill. But a fast gas fill can result in the wrong percentage of gases. Oxygen and helium must be decanted slowly and blended evenly.

It pays to pay an expert to do the job. Furthermore, there's more to mixing gas than simply putting in the gas. The mixing station must maintain certain grades of purity in the gas it dispenses, must fill tanks from an oil-free compressor or have adequate filtration, must change costly filters frequently in order to maintain purity, must send gas samples to a testing facility to be tested for contamination, must build and maintain the blending system, must pay demurrage on gas storage cylinders or purchase them outright, and must amortize the capital expenditures that went into purchasing the tubing, hoses, valves, connectors, and compressors and pumps.

Keep this in mind when you see the bill.

Slow Mixers - Shaken not Stirred

Just because the tanks are filled doesn't mean that the gases are mixed. Gases of different densities tend to stratify, to maintain their autonomy and personal space. Oxygen and helium must be forced together like reluctant lovers. This might take a few hours.

For this reason, some technical shops with mixing stations don't like to release the tanks to their customers until the gas can mix and be analyzed. Gas in a tank that is too carefully handled might never mix. The way to make the gases mix is to shake the tanks around. Single tanks too heavy and awkward to lift can be rolled every few minutes, and doubles can be flipped.

The mixing process can be accelerated by putting the tanks in the sun, where the heat creates convection currents that stir the gases into action like Macbeth's three witches.

Ted Green's method is to slam the air on top of the oxygen and helium, then let the tank settle and top it off. The sudden injection of air breaks up the stratification and forces the gases to mix.

Analyzing for Oxygen

After a breathing gas is blended it *must* be analyzed. This is commonly done with an oxygen analyzer. One can be purchased for a few hundred dollars. While this may seem like an unnecessary expense for an individual who purchases his gas from a reputable dive shop, which verifies a mix before it leaves the store, it's a comfort to many to be able to check their own.

One of the accepted rules of technical diving is: analyze your own gas.

This doesn't mean that you have to buy your own analyzer. When you purchase gas from a dive shop, the blending technician will place a tag on your tanks which specifies the percentage of oxygen in the mix. It's a simple matter for you to observe the gas being analyzed, or to analyze it yourself with the shop's analyzer. Get familiar with the process.

It pays to double-check because anyone can make a mistake - and this kind of mistake can be fatal or worse. If you're smart, you'll follow my recommendations in the section below on getting a second opinion.

The only gas in a mix that is verified by analysis is oxygen. The remainder is inert gas, called diluent. Helium analyzers exist but they cost thousands of dollars. Analyzing for helium is generally superfluous since the most immediate concern is that the partial pressure of oxygen will not be toxic.

Simple subtraction will tell you the percentage of inert gas in heliox. Whatever isn't oxygen is helium. Likewise, heliair contains only two essential ingredients: helium and air. The percentage of gases in air is fixed and known.

In trimix, the mathematics is more complicated because the diluent is a combination of two inert gases. Testing for oxygen will yield the percentage of inert gas, but not the proportion of nitrogen to helium. For that determination you need to know the various partial pressures that were put into the tank, then subtract the result from the total pressure.

In actual practice, the helium fraction can be off by as much as 5% without materially affecting the decompression profile. Let me state again that the primary concern in breathing mixed gas is the partial pressure of oxygen. Offsetting narcosis is secondary.

Caring for your Analyzer

Oxygen analyzers are simple to operate: pass gas through a hose to the oxygen sensor and read the digital display. But like any sensitive instrument it must be employed properly in order to furnish accurate readings. Two components need special attention and flow characteristics must be fully understood. First, the components.

Portable units are powered by dry cells that must be fully charged. Reduced voltage from a weak battery can make the readout "wobble." That is, the numbers might jump from high to low without ever settling out. At first you might think the electronics have gone screwy, but replacing the battery usually puts everything aright.

The guts of the unit is the galvanic oxygen sensor, sometimes called a fuel cell. The sensor contains an electrolyte which reacts with oxygen that diffuses through a membrane. By means of electrochemical reduction and oxidation, the electrolyte produces a weak electrical current which is proportional to the partial pressure of oxygen in the gas sample. This current is then amplified to drive the display.

Because the electrolyte is consumed in the process of oxidation, the sensor has a limited life span beyond which it no longer detects the presence of oxygen. When the sensor goes bad, the readout generally goes blank, although sometimes the readout might fluctuate without ever settling out. (The readout will also go blank if the battery is completely dead.)

Except for the cost of a new sensor, this is no big deal. The sensor can be changed as quickly as changing the batteries in a flashlight. When in doubt about the cause of a blank or erratic display, change the batteries first. If that doesn't solve the problem then you probably need a new sensor.

In a sense, the sensor works all the time: that is, all the time it's exposed to an oxygen environment. Even when the analyzer is not switched on, a chemical reaction is taking place with the electrolyte as atmospheric oxygen consumes the reagent. This is what limits the sensor's longevity.

Replacement sensors are stored and delivered in sealed foil packets which contain pure nitrogen. The sensor's reagent doesn't react with nitrogen, only with oxygen, so the sensor doesn't begin deteriorating until the packet is opened (and the sensor installed).

Technically, a spare sensor kept in its original sealed packet has a shelf life of infinity. Like a newborn child, life doesn't count until the electrolyte draws its first breath of air. But in actuality this isn't the case. Some chemical degradation occurs no matter what. Since you can't predict exactly when a sensor will go bad, and since sensors are expensive and lose some efficacy during storage, the question that arises is: when should you buy a replacement?

You might say: don't get a new sensor till you need it. This admonition violates the technical diving motto: always have a back-up. If the sensor takes its last breath during an overnight boat trip or a multiday cave exploration, you might find yourself, in the unforgettable words of my ninth grade Latin teacher, *sine ulla pedulla* - up the creek without a paddle.

The best advice in this regard is to accept certain expenditures as unavoidable investments and insurance against untimely failure. The very nature of technical diving demands equipment purchases and maintenance costs in excess of its recreational counterpart: givens that must be taken as a matter of course.

Optimize analyzer upkeep by purchasing a new sensor some months before the life expectancy of the old one is due to expire. Then amortize the cost over the span of the sensor's serviceable lifetime.

When the analyzer is not in use - between trips or during the off season - you can suspend or retard the aging process by removing the sensor and storing it in a vessel filled with nitrogen. This isn't practical. But a stopgap solution is to stow the entire analyzer in an airtight container such as a plastic camera case or gadget box that seals with a gasket, in which the oxygen available for chemical reaction is limited and won't be renewed till you open the container. Foam inserts can be cut out so the analyzer and attendant parts fit snugly and are protected from physical abuse.

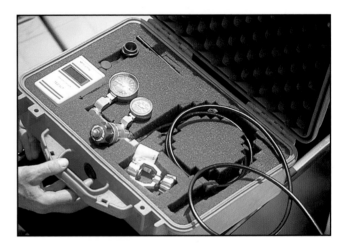

Calibration Techniques

Each time you switch on the analyzer you should check the calibration, test the gas, then verify the calibration. If the calibration is off, repeat the process. If the calibration is still off, check the battery and sensor.

In the (technical diving) field, the calibration is usually checked by swishing the lead of the tube or the gas collector through the air, in which the concentration of oxygen is known, and waiting for the readout to settle on 20.9%. That's the partial pressure of oxygen in air at sea level (but not at higher elevations). It might take a minute or so for the reading to stabilize.

If the readout registers high or low, the unit is either out of calibration or you're near someplace where the air is substandard or full of pollutants, such as a freeway at rush hour, a refinery at lunch time, or a political debate. If the calibration is off, rotate the manual calibration dial until the readout registers correctly.

In the instruction manuals that come with most analyzers, the manufacturers recommend that a more accurate reading can be attained by calibrating the sensor in a pure oxygen environment or from oxygen siphoned off an oxygen cylinder. This recommendation is made because analyzers are ordinarily marketed to doctors and hospitals for use in operating rooms, where the oxygen to be analyzed is pure.

The accuracy of an analyzer is greater when the sensor is calibrated close to the concentration of oxygen in the gas that you expect to analyze. For technical diving applications, this means that instead of calibrating the sensor at the high end of the scale, it should be calibrated at the lower end, in the vicinity of air.

Room or outdoor air will suffice, but compressed air is even better. The reason for this is explained in the following section.

Go with the Flow

As important as the concentration of oxygen in a gas sample is the flow rate. It's important to understand that the analyzer measures *pressure* within preset parameters according to instructions in the operating manual. Gas flowing too fast into the sensor will give a reading that's too high, while slow moving gas will give a reading that is low. The rate of flow must be controlled in order to obtain an accurate reading. This is equivalent to an automobile's gas gauge whose needle is affected by a steep hill; only on level ground can you be certain how much fuel remains in the tank.

If your analyzer doesn't come with a flow restricter, get one. By merely holding the collector cup or tube end against the tank valve orifice, then opening the valve, you can't be sure that the flow rate is optimal or that atmospheric air isn't being sucked in around the lip by the venturi effect. Contaminating the test gas with air will yield an erroneous reading.

A flow restricter can be as expensive and sophisticated as a pressure reduction device complete with tank yoke, adjustment knob, and calibrated scale; or as cheap and simplistic as a metal block that quick-connects to a BC connector or

that screws into a low-pressure port, and that restricts the flow of gas by means of a metered orifice or calibrated pinhole whose diameter determines the speed at which gas passes through. Either adaptor is acceptable as a pressure reduction method.

Analyzing Accuracy

One way to calibrate the analyzer and correlate values is to first analyze a cylinder of pressurized air, then analyze the gas in the tank in question by using the same flow rate. Assuming that the battery is good and the sensor is in working order, a reliable comparative reading can be obtained.

The accuracy of analyzers lies in the range of plus or minus 1-2%. For a deco gas, being off by 2% isn't bad if you only want to verify which tank has air and which has nitrox-32. But if you're blending bottom mix that's supposed to be heliox-12, and you can't determine if the final blend is heliox-10 or heliox-14, you could be looking for trouble.

The percentage that an analyzer is off of true relates to the full range of the oxygen scale, not to the percentage of oxygen in the mix. For example, a divergence of 1% means that the oxygen fraction in heliox-12 could be off in either direction, and that the actual blend could be anywhere between heliox-11 and heliox-13. By comparison, nitrox-50 could be either nitrox-49 or nitrox-51.

This amount of unknown deviation might be acceptable if the oxygen content falls within the tolerance range of toxicity and if the decompression profile provides enough latitude with respect to decompression conservatism. A 2% deviation might not offer enough leeway or margin of safety in a bottom mix, but could be adequate in a deco gas.

In the heliox-12 example given above, the margin of error in the analyzer is greatly reduced because of the proximity of the fraction of oxygen in the calibration gas (air) as compared to the fraction of oxygen in the mix.

For truly accurate calibration that spans the full scale (from 0% to 100%) conduct a "two point linearity check." To do this, analyze two gases with known concentrations of oxygen: one in the high range (90-100%) and one in the low range (say, atmospheric air). Both should check out correctly. If not, the sensor may be going stale.

Patience is a Virtue

It takes a few moments for an analyzer to come down after testing a gas with a high ppO₂, in the range of nitrox-70 or pure oxygen. You might think to rush the process by blowing air into the collection tube. *Do not do this!* The carbon dioxide in exhaled breath is not good for the sensor.

Let the readout settle on its own, or accelerate the subsidence by swishing the tube collector assembly through the air.

Getting High

Although an analyzer measures oxygen by means of partial pressure, it works just as well at altitude as it does at sea level despite reduced atmospheric pressure. This is because analyzers, as noted above, operate on a linear scale; and because the reduction of total pressure doesn't change the proportions of constituent gases within a mix.

The partial pressure of oxygen in air at 5,000 feet is less than 21%, but the percentage remains the same.

If you're mixing gas at high elevations, or checking tanks for oxygen content prior to plunging into a high altitude lake, simply switch on the analyzer on the spot, let the readout settle, then adjust the reading to 20.9%. Once calibrated for ambient pressure, the analyzer will determine the concentration of oxygen from its new set point.

In the Final Analysis

An analyzer will respond to changes in barometric pressure. As noted in the preceding section on high altitude testing, as long as you calibrate the readout whenever you switch on the analyzer, the sensor will operate from the established set point and will yield accurate readings.

Analyzers are also affected by temperature and humidity. Thermal protection is provided for the range between the freezing point of water and room temperature. Avoid testing in extreme cold. Also, refrain from excessive handling of the sensor. Body heat can raise the temperature of the sensing electrode and disturb the thermal equilibrium. "Handle with kid gloves" might be an appropriate aphorism here.

Humidity within a gas mix will reduce the percentage of oxygen. This shouldn't be a problem for breathing gases if adequate filtration is employed in the pumping process. Water separators upstream of the intake ensure that gas is delivered dry.

On the other hand, atmospheric humidity can cause moisture to condense on the sensor membrane. Microdroplets create a patchwork film or partial barrier which blocks the free passage of oxygen. This will lead to an erroneous reading.

Care should be taken when analyzing for oxygen on a boat at sea, where moisture is fed to the atmosphere by evaporation from surface water. Comparative readings taken on land and at sea might therefore yield slightly different results.

Get a Second Opinion

Analysts will love me for pressing the point that you should always have a double analysis. Redundancy is paramount in all aspects of technical diving and assumes greater significance when it comes to gas analysis. A couple of percent off in the concentration of oxygen can cause serious trouble, from oxygen toxicity to DCI. Don't chance it.

The prudent diver will analyze his gas twice and from two different sources. By this I don't mean to imply that you need to own two oxygen analyzers. If you test your gas with the analyzer at the shop that dispenses your mix, then check it again yourself at home or at the dive site, one verifies the other. If you're making home-brew (see the next chapter) a second opinion is even more important.

Still, you don't need to own two analyzers as long as you have access to a back-up. People active in the technical diving community seldom operate alone. If several members in a group have analyzers, use them cooperatively.

In case of a difference, which analysis do you believe? The correct answer is neither. Recalibrate both analyzers then test the gas again. If the analyzers are in working order and calibrated correctly, they cannot disagree by any significant amount. If you continue to get erroneous readings, the gas mix must be wrong.

Gas Storage

Storing helium mixes can pose a problem. Not because the gas will go bad (it won't) but because the unused gas occupies tanks that cannot be otherwise employed. Helium is so expensive that you can't justify blowing it off if a dive is canceled. And the percentage of oxygen in the mix imposes limitations on the depths at which the particular blend can be breathed. It's almost cost effective to buy more tanks.

The situation is more complicated for wreck divers than for cave divers. Boat trips are exceedingly weather dependent, and a dive to the same wreck or to a similar depth often can't be fitted into a boat schedule that's already filled.

There is no easy solution to this dilemma. I have no miracle cure to offer. I can only state that an activity that is already expensive and equipment intensive can be more so.

In the early days of technical diving it was believed that helium could leak out of a tank around the valve threads, thereby altering the mix by increasing the percentage of oxygen. I have not found this to be true. If your tank springs a leak through the o-ring, the volume of the gas in the tank will be reduced, but the proportion of gases in the blend will remain the same. Gases are good mixers: once mixed, they stay mixed.

I have stored heliox-12 for nearly a year without losing a single psi.

Fill 'er Up

Hopefully, after a mixed-gas dive you'll have some gas remaining in your tanks. If the blend contains a high percentage of helium you'll naturally want to save what's left in order to reduce the cost of a refill. Nothing could be easier than topping off from a bank of pre-mix that's the same as what you've already got.

But most gas blends are specially mixed with specific percentages of oxygen and diluent. This means that each gas is added separately by means of partial pressure. Equations exist for calculating the quantity of each gas to be added in order to reach the desired blend, but the applications are limited.

For heliox, as long as the total pressure is low enough you can add oxygen up to 2,400 psi, then fill to capacity with helium. You'll know if the result is accurate when you analyze the gas.

Trimix is more difficult to remake because there are two unknowns in the equation. First add oxygen, next add helium, then fill up the rest with air. If the partial pressures were measured precisely you *should* end up with the targeted mix. But you can analyze only for oxygen. Without a

helium analyzer there's no way to calculate the exact proportion of inert gases in the diluent.

The element of uncertainty is exaggerated with each succeeding fill as additional error is introduced. You then run the risk of violating the decompression requirement.

The general rule of thumb is to top off trimix only once. Then dump the gas and begin from scratch.

The Deep Frontier

Is helium diving worth all the time and trouble? Undeniably.

Is this kind of diving for everyone? Not by a long shot.

Strict discipline and attention to detail are prerequisites for safe conduct. Helium diving is not a stunt, but a tool to accomplish what cannot be done on air.

At medium depths, the addition of helium can bolster experienced divers who have an unusually low tolerance to nitrogen narcosis. I have friends who are zonked out of their gourds at 160 fsw, yet who are competent and capable otherwise. What helium can do for them is to alleviate the symptoms of narcosis and let them operate at their full potential.

This doesn't mean that anyone can strap on a set of doubles filled with mixed gas and plunk down to 300 fsw. The margins for error are great, and can be overcome only with training, careful planning, the proper equipment, the right gases, conservative decompression schedules, and a healthy respect for the challenges involved - tempered by long-time experience in going deep.

Technical diving offers no guarantees for success. The underwater world is a fickle place.

Perhaps the best approach is a cooperative effort in which the overall costs are shared by a group of divers all seeking to explore a new frontier.

For some, the challenge lies not only in the performance of the dive, or in the proficiency of the system, but in acquiring the esoteric know-how needed to prepare for such a venture. A position on the leading edge of diving technology is a lure that can become a goal in itself.

And as an added benefit, mixed gas will enable you to go literally where no one has gone before.

That makes it all worthwhile.

EXPEDITION DIVING

"Expedition" is defined in the dictionary as "a journey undertaken by an organized group of people with a definite objective."

By broad interpretation, I suppose the meaning of expedition could encompass such diverse activities as a cross-country political campaign to hoodwink voters, a vacation flight to Hawaii for a lei, or a jaunt to Las Vegas to gamble and get drunk. Somehow this seems sacrilegious.

All too often the word "expedition" is used as hype or as an advertising ploy. It describes an undertaking whose goal the promoters or participants would like to be perceived as more lofty than it is. In such cases, it is hoped that the emphasis on "expedition" will lend legitimacy or genuineness to an otherwise obscure enterprise, or to exalt a project whose pursuit possesses little value except to those who are engaged in it. This kind of expedition is generally a money-making operation.

In the strictest sense, expedition could describe an overnight climb to the summit of a snow-covered mountain in winter, a one-week backpacking adventure through a secluded desert canyon, a month-long wilderness paddling excursion by canoe, or a multiday oceanic enterprise to discover and dive on long-lost shipwrecks. I've been on many such outings and have always thought of them simply as "trips." They were done purely for my own enjoyment and I make no pretensions otherwise.

In my mind, the classic image of an expedition is exemplified by such exploits as the mapping of western America by Lewis and Clark, Darwin's voyage to the Galapagos Islands, Burton's and Speke's quest for the headwaters of the Nile, Stanley's search for Livingstone, Shackleton's Antarctic attainments, and the conquest of Everest by Hillary and Tenzing.

It seems to me that the perception of "expedition" is dominated by the notion of geographical discovery, travels to distant lands, and the exploration of the far corners of the world where people do not live or where civilization has not yet encroached - with the implication being that the explorers are left to their own devices without any hope of outside rescue should their plans go awry or should natural catastrophe occur. A certain amount of risk is involved. (Oddly, no one seems to think of a mission to the Moon as an expedition, although it stands undoubtedly as the "loftiest" and most ambitious effort of exploration in all of human history.)

A Modern Interpretation

The scale and scope of present-day ventures pale by comparison with expeditions of the past, some of which were years in the accomplishment. Nevertheless, it cannot be denied that forms of exploration still exist and that extraordinary endeavors are currently undertaken by enterprising people with profound destinations in mind. These exploits require planning, organization, dedication, and the acquisition of mental and physical skills in order to successfully achieve their aims.

The complexities of present-day society impose constraints on people that pioneers of the past didn't have: education, career, family obligations, and financial responsibilities such as mortgages and insurance payments, to mention a few. These are life-long commitments that cannot be conveniently suspended. Nine-to-five jobs with only evenings and weekends free thwart a person's eligibility for lengthy leaves of absence.

But society hasn't eliminated - nor will it ever eliminate - the individual's innate desire for excitement, exploration, and the challenge of the unknown. These qualities are part of our biological heritage. So while a midnight stroll through the wild side of town may provide a momentary surge of adrenaline, surviving such a promenade is not likely to satisfy the deep-seated yearnings that drive people to test their personal fortitude, to realize their full potential, and to seek out new frontiers.

I submit that "expedition" must be redefined in modern perspective, that old fashioned conceptions need updating.

Which brings us to technical diving. While it's figuratively true that the world has gotten smaller, it has also, metaphorically speaking, gone deeper into the soul of exploration. Many of today's deep-diving initiatives are extremely well organized, boast bona fide objectives of geographical discovery, incur risk, and are remote in the sense that they are conducted far away from metropolitan comfort, off the normal routes of convenient supply, and in depths where deliverance is totally out of the question.

Furthermore, some of these "trips" reach true expeditionary proportions in the amount of preparation required, the logistics involved, the number of people included, and the total time invested between initiation and conclusion.

Two well-known examples are the 1987 Wakulla Spring project, which extended the known tunnel system of the submerged Florida cave into previously unreached territory, and the exploration of the *Lusitania* in 1994, an expedition in which the author was one of twelve participating divers. These "expeditions" demanded several weeks on site, to say nothing of mobilization and demobilization.

A book was written about the Wakulla Spring project. It must be read in order to fully appreciate the complexities of a group, mixed-gas cave exploration that was years in the planning and

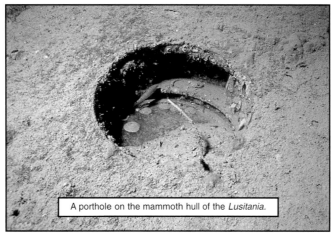

A porthole on the mammoth hull of the *Lusitania*.

months in the execution.

I will discuss some of the salient technical features of the *Lusitania* dive trip after I lay some groundwork.

Small Scale Assault

Perhaps as demonstrative as diving Wakulla Spring and the *Lusitania* is a new form of exploration in which a project, such as the extended penetration of a distant tunnel or the survey and identification of a newly discovered wreck, is stretched out over an entire season or, in some cases, goes on for years. In such situations the days of diving do not occur consecutively. Instead, dives may be conducted only on weekends as jobs and social duties allow.

This method of approach is more in keeping with community lifestyle. And since expeditionary measures are incorporated for the duration of an exploratory project, although admittedly in a sporadic fashion, I submit further that the modern meaning of "expedition" must be expanded to fit the way in which perilous exploration is now pursued.

In mountaineering, the final thrust of a climbing party to attain the summit of a peak is called an assault. The word is used without implication of violent attack in the military or judicial sense. It means instead that there are obstacles to overcome in reaching the prime objective. By extrapolation, small scale diving operations can be considered as assaults: progressive penetration past the farthest known point in a deep submerged cave, or overnight boat trips to offshore sites.

In my specialty - the search for and exploration of undiscovered shipwrecks - the logistics involved in planning weekend assaults can be overwhelming. I have literally tramped the docks in some rural areas, where charter services were unavailable, seeking boats to take me to wrecks that I knew lay off the coast. I've also chartered boats to travel hundreds of miles from their home marinas, in order to reach wrecks off rocks, islands, and deserted coastal zones where wreck diving had not yet arrived. I've hired fishing boats, lobster boats, shrimp boats, and inflatables: anything afloat that would take me to a wreck site.

These types of assaults can be costly - not only in time but in money. Chartering a boat implies gathering a group of like-minded divers to share the costs. If you can't convince enough people to subsidize your interests, you either pay extra for the privilege or overcharge those who lend their support. Then you have to deal with the personalities involved, especially when a trip fails to achieve anticipated goals and the divers are left unsatisfied.

All this and more is part of being a trip leader, or expedition leader. Before you can undertake trips of greater proportion, it is necessary to become proficient in leading small scale assaults which those who are a step ahead of you on the treadmill of experience take for granted.

Join expeditionary type endeavors as a participant, observe how they are run, gain some insights, and take the torch when it's passed.

Do-it-Yourself User's Guide

Justifications notwithstanding, this chapter deals with ways to implement deep diving operations that are a cut above - or below - typical weekend dive trips, while at the same time enlightening the reader with such esoteric knowledge as how to mix gas, how to choose the best blend for the depth, and what hardware is needed.

The revelation of this information is necessitated by the fact that, despite the proliferation of technical dive shops, not everyone lives conveniently close to facilities where mixed gas is dispensed; or the hours of operation may be unfavorably restrictive; or the areas of activity may be located in rural areas or far at sea.

Even if you begin a trip with your tanks filled with gas, once you reach the dive site you may find it impractical to return to the shop for a refill between dives. If you don't own or can't carry enough tanks to store all the gas you'll need for the trip, you might have to rent storage bottles or find some other means of having an adequate supply. Or you might have to mix your gas on site.

I've dived in the Great Lakes where I had to drive for half a day just to get *air*. Divers who explore submerged caves in remote parts of the country, far from the mainstay of the technical diving community, might find it more cost effective to invest in some gas transfer equipment so they can blend their own breathing mix.

Self-sufficiency may be expensive, but sometimes it's the only way to go. Patronize a technical dive shop if at all feasible. Otherwise, read on.

The "Hot" Mix Concept

One time I planned to dive on the battleship *Ostfriesland*, at 380 fsw. My tanks were filled with gas mixed by a technical dive shop the week before the dive. When the trip was called off at sea due to mechanical problems with the boat, I was stuck with a set of doubles and a single side-slung full of heliox-12. I was all gassed up with nowhere to go.

The trip couldn't be rescheduled for months because the only boat that could take us to the site was booked far into autumn. Even then there was no guarantee that the trip would eventuate before weather forced the diving season to wind down. In the meantime my tanks were tied up. The heliox represented a sizable investment so I wasn't about to dump it. Then came the opportunity to dive the *Norness*, which was 100 fsw shallower.

I could dive to 280 fsw on heliox-12 but it seemed like a waste of a good mix. Furthermore, the potential existed to make three dives on the *Norness* because we planned to stay grappled in the wreck overnight. What to do?

I showed up on the boat with my three tanks filled with heliox-12, plus an empty set of doubles and an empty single. I connected the two sets of doubles by means of a transfill hose, opened both valves, and let the gas flow until the pressure in both sets equalized. This gave me two sets of doubles that were half-filled with heliox-12.

Using the boat's compressor, I topped off both sets of doubles with air. I tested the resultant blend with an oxygen analyzer. I now had two sets of doubles filled with trimix-16.5/44. With a laptop computer program I cut a decompression schedule based upon the inert gas content and breathing the same deco gases that I had planned to breathe on the *Ostfriesland*. The resulting profile was not as efficient as it could have been with a better choice of deco gases, but it made practical use of what was already available.

Afterward, I equalized my two single tanks

and topped them off with air. This yielded the same mix. I removed the bands from a set of doubles, slipped them over the singles, and rigged for the third descent by diving "splits": that is, instead of using a tank manifold with an isolator valve, I alternated regulators and breathed one tank down to 1,000 psi, then switched to the regulator on the other tank. Not my preference, but any port in a storm as they say.

What I did as an expedience out of economic necessity, others on the boat that day observed and later established as methodology. You might say that I provided the kernel of an idea while others popped the corn. Thus was born the intentional "hot" mix.

Utilizing Hot Mixes

A hot mix is one which is purposely blended with higher percentages of oxygen and helium than those intended to be breathed on the actual dive, for the express purpose of later creating an appropriate breathing mix by adding air. "High" mix would be a more accurate description since "hot" implies radioactivity or scorching temperatures, but the term hot mix is the one in prevalent use.

Take for instance an *Andrea Doria* trip on which five dives are planned over a three day period. Let's say you'd like to breathe trimix-17/25 on the wreck. You bring three sets of doubles on the boat. One set contains trimix-17/25, the other two sets contain trimix-13/50. (At a depth of 240 fsw trimix-13/50 can be breathed but is less than optimal.)

For the first dive you use the set that contains trimix-17/25. After the dive you drain those doubles completely, then connect them by means of a transfill hose to one set of tanks containing the hot mix. Equalize the pressure, top off each set with air, analyze for oxygen, and you should end up with trimix-17/25 in both sets.

The helium content is *decreased* by half when it's split between two tanks. The oxygen content will change by half the difference of the oxygen content between the two blends (assuming tanks of the same size and fill pressure). In the hot mix example above, adding the two fractions of oxygen (13% in the trimix and 21% in air) equals 34%. By dividing by 2 we get 17%.

Despite the simplicity of the mathematics and the minuscule margin for error, always analyze for oxygen. In mixed-gas diving never take anything for granted.

After making two more dives, repeat the procedure with the second set of doubles that are filled with hot mix, and you've got your gas for the final two dives.

• Hot mixes must be labeled as such so you don't breathe them accidentally.
• If you're using doubles with an isolator manifold, it is vitally important that the isolator valve be open during the transfill and topping off procedures.
• After transfilling and topping with air, make sure the gas is thoroughly mixed. Adding air by means of a compressor is slow, so you don't get the mixing effect that you get when air is jolted into a tank from a high-pressure bank. To accelerate the mixing process, roll singles and flip doubles.
• Verify oxygen content with an analyzer.
• Note that recipient tanks in the transfill process do not need to be cleaned for oxygen service. The percentage of oxygen in any helium breathing mix is less than 40%.

By utilizing hot mixes in the example above, the number of sets of doubles needed for a five-dive trip is reduced from five to three. This not only reduces the number of tanks you need to own, but it makes more space on the boat.

Some deco gases can be hot mixed, too, if the desired percentage of oxygen is less than 60%.

Home-Brew

Any gas mix that is blended without professional assistance is called a "home-brew." Technical dive shops and commercial gas suppliers tend to ridicule home-brews because they - the shops and suppliers - are left out of the profiteering loop. But there's no constitutional prohibition against making this kind of brew in the basement or garage. Nor is a home-brew by definition any less safe for breathing than a commercially blended gas provided by a mixed-gas merchandiser.

Precautions must be taken. It goes without saying that home-brews must be blended slowly and carefully, then verified for accuracy by analyzing for oxygen content. Where necessary, equipment must be cleaned for oxygen service. Do *not* presume that oxygen cleaning rules apply only to commercial gas merchants. A flashover results from a law of physics, not a law of government; it can't be repealed and variances are never allowed.

The Cascade System

The simplest way to blend gas is to cascade it. To cascade means "to fall from one level to another."

The basic cascade system consists of a gas supply and a transfill hose with the appropriate thread adaptor. Connect the transfill hose to both the gas cylinder and an empty scuba tank, open both valves, and let the gas flow. The gas "falls" from the tank whose pressure is higher into the tank whose pressure is lower.

In essence, bleeding helium from a storage cylinder into a scuba tank is only slightly more complicated than filling a party balloon.

That's the short version. Here are the details and complications.

Fractional Gas Blending

Start out by getting your terminology correct. In the field you're likely to hear divers talk about "blending by partial pressures" or "partial pressure blending." This phrase is inaccurate and misleading. It likely stems from a misunderstanding of Dalton's Law, which states "the partial pressure of each gas in a mixture is proportional to the relative amount of that gas in the mixture."

Dalton's Law refers to the force that a gas exerts. It does not mean to imply that a cylinder is filled with a bunch of partial pressures as if they were grapes in a bowl. The pressures of individual gases are partial only in relation to the whole. A quantity of gas that is introduced into a cylinder represents the *entire* pressure (or force) of that gas, not its partial pressure.

What are actually introduced into a cylinder during the gas blending process are fractional components. In a laboratory these fractional components can be weighed or counted in moles. In large-scale industrial applications a gas is quantified by cubic feet. Divers tend to think in terms of

Silver cylinders of helium and green cylinders of oxygen are clustered on the ground during mixing operations in the marina prior to a trip to the *Andrea Doria*.

pressure rather than volume because that is what tank gauges measure. Volume is an interpretation. All of which relates to the sections in the chapter on "Gear in Depth": "Tank Pressure versus Tank Volume" and "Pressure Gauges are not Volume Gauges."

It is more accurate to think in terms of "fractional gas blending" than "partial pressure blending." That is, the fractional component (such as oxygen or helium) of a gas mixture is introduced into a cylinder, then another fractional component (such as air) is introduced. That these fractional components also exert a force or pressure is another matter entirely.

My objection may seem like an exercise in semantics, and to a certain extent I suppose it is. But given the choice of being precise or vague, I prefer precision.

In practice, since other methods of blending gas are not generally employed in technical diving (although other methods may be used in scientific and industrial applications) you can simply use the term "gas blending."

Thread Adapters

Gas storage cylinders are fitted with valves similar in design to a scuba tank valve. The purpose of a valve is to regulate the flow of gas. The CGA has designated specific nut and nipple configurations to the valves that regulate the flow of each elementary gas and certain gaseous compounds.

For example, an oxygen cylinder is equipped with a valve design which accepts only an oxygen compatible thread. You can't screw an acetylene regulator into an oxygen valve by mistake. This straightforward measure prevents a welder from connecting his hoses in reverse order when he needs to blend oxygen and acetylene in a torch to provide a flame for cutting metal.

For technical diving applications, the valves generally encountered are those intended for use with air, oxygen (CGA 540), and helium and argon (CGA 580 for both). Thus a transfill hose may have air threads on both ends in order to equalize the pressure between two scuba tanks, as described above; or it may have an air thread on one end and a different thread on the other end, depending upon which gas you intend to cascade.

This doesn't mean that you need to buy a separate transfill hose for each gas. You can obtain metal thread adapters that connect to an air thread on one end and a gas thread on the other end. One hose with two adapters is the universal expedient, as long as the hose is cleaned for oxygen service. Or you might choose to have two hoses: one for air and oxygen, another for air and helium and argon.

Regulating Gas Flow

Opening the valve on a gas storage cylinder lets out gas. If the valve is connected to a transfill hose, the gas will flow through the hose into the recipient tank.

I have seen admonitions that the oxygen tank valve was not designed for throttling purposes (controlling the flow rate) but was intended only as a gate (for opening and closing). By extension, the warning might also apply to tank valves used for other gases as well. Such warnings recommended the use of a flow restriction device. This is nothing more than a coupling with a small hole that limits the flow of gas by virtue of the diameter of the orifice.

The purpose of regulating the flow rate is twofold: to obtain the proper amount of gas, and to reduce friction against the valve seat. Friction can conceivably cause ignition in an oxygen environment. Over time, friction can erode the seat and promote wear.

In the field, however, many home-brewers dispense with fancy accessories and make do with the simple valve that is supplied with the storage cylinder. They regulate flow by careful manipulation.

I suspect that for industrial applications special regulators and flow restricters might lower maintenance and replacement costs. But the trifling usage entailed by technical divers doesn't seem to warrant the investment. I will not go as far as to state that a pressure-reducing regulator is unnecessary, or that it is not a good idea. But I will state that no technical divers that I ever met employed them when making home-brew. They cracked the valve a bit and let the sound of hissing gas be their guide to controlling flow.

There are recommendations - and then there's home-brew.

The Cascade Pressure Gauge

Because most breathing blends are mixed by means of pressure, an accurate pressure gauge is essential - perhaps even vital. A submersible pressure gauge is inadequate for the task. When 640 psi of oxygen is needed in an empty tank in order to achieve the precise percentage of oxygen in the final mix, a reading of 600 psi or 700 psi will not suffice.

You can squeak by with a stock pressure gauge such as the kind that is furnished with an oxygen regulator purchased from a hardware store. But the dial is so small that it's difficult if not impossible to read with exactitude.

Much better for technical diving purposes is a gauge the diameter of a small dinner plate. The large dial permits smaller increment sizes which allow for greater accuracy - you can tweak the needle around the dial.

Alternatively, although they are more expensive, digital gauges are also more accurate, infinitely easier to read, and can be zeroed in the field without having to be returned to the manufacturer for recalibration.

Pressure Gauge Calibration

As I've already noted, there's a difference between precision and accuracy. A pressure gauge not will give an accurate reading if it isn't calibrated with precision. Gauges are calibrated at the factory, so a new one should read accurately right out of the box. But with continued usage over time they tend to lose their calibration.

This happens so slowly that you might not be aware of it until you find your oxygen analyses off their intended mark, or discover that the final pressure shown on your submersible gauge disagrees. It's a matter of little difference whether a tank filled with air gauges out at 3,500 psi or 3,550 psi. The 50 psi is merely a matter of content. But in blended gases such a difference may be crucial.

If there's a saving grace in the rising error in gauge calibration, it is that more often than not the percentage of calibration deviates consistently for the full range of the scale: that is, it will be off the same amount at 500 psi as it is at 3,500 psi. This means that the fractions of cascaded gases will be proportionate to each other so that their fractional components will not be affected; only the total pressure will be erroneous.

But suppose you cascade only the oxygen for

a nitrox blend, then have the tanks topped off elsewhere with air? Now two gauges have been used. If they are calibrated differently, the fraction of oxygen in the final blend will not be what you wanted.

A gauge can also incur spot miscalibrations. In this case, it might be off more at 1,000 psi than it is at 2,000 psi. This kind of error may be difficult to prove.

I'm not trying to scare any one. I just want you to be aware that pressure gauges are not infallible. They should be returned periodically to the manufacturer for recalibration. And if you drop the gauge or otherwise jar the working mechanism, you would be wise to have the instrument checked for accuracy.

Pass Gas Slowly

A fast fill is often called a hot fill. In this case "hot" is meant in the literal sense. The rapid expansion of a gas raises the temperature of the gas, so that a tank that is filled too quickly feels warm to the touch.

Everyone knows that raising the temperature of a gas also increases its pressure. As the gas in a tank cools, the pressure is reduced, so a tank that is gauged full at the shop might gauge a few hundred psi less at home or at the dive site. The result is a short fill.

There is no room for short fills in gas blending because it upsets the delicate balance of the blend. Gases must be cascaded slowly in order to check expansion, and so the gauge registers the actual pressure of the gas in the tank and not a temporarily expanded pressure.

Take the example given above, in which 640 psi of oxygen is required in order to achieve a specific fraction of oxygen in the final mix. If you cascade the oxygen too quickly, and if you don't give the heated oxygen time to "settle out" before topping off with air or before cascading helium on top of it, the percentage of oxygen in the final mix will be less than that anticipated.

You will ascertain this when you analyze for oxygen, but you won't be able to correct it because you can't add more oxygen at the end of the blending process - oxygen should not be pressurized higher than 2,400 psi, and the final pressure is too great for adding oxygen. This means that you're stuck with a blend in which the fraction of oxygen may be insufficient for your needs.

If the oxygen content is only slightly off, you might be able to reprogram your decompression profile. Otherwise you'll have to dump the gas and start over.

Another reason for decanting oxygen slowly is to prevent ignition or flashover which can result from high-speed friction. The recommended speed limit for passing oxygen is 60 psi per minute.

Helium is an inert gas and will not ignite, but the caution about overheating still applies.

Maintain your Cool

Ambient temperature also affects the pressure of a gas. On expeditions, gases are seldom blended in the comfort of a temperature controlled room. Gas may be blended on the bed of a truck, on the ground in a forest, on the deck of a boat, or at the dock.

Out-of-doors gas blending can be tricky. In addition to swatting flies and wiping sweat off your brow, you have to contend with keeping recipient tanks cool. It isn't always possible to have a water-filled container to act as a heat sink, such as those found in dive shop filling stations. Oftentimes, the best you can do is play a stream of water from a hose over the tanks. Or wrap the tanks in wet towels to take advantage of evaporative cooling.

Do whatever you can to lower the temperature of the tanks. And not just the recipient tanks, but the gas storage cylinders as well. It stands to reason that a storage cylinder left in the sun will acquire heat from infrared radiation, and that this heat will increase the pressure of the gas within the cylinder. Store cylinders in the shade if possible and do your blending there as well. The last thing you want is a gas blend with sunstroke.

Consider this scenario from a recent deep wreck exploration week that consisted of a series of overnight boat trips separated by dock days when gases were mixed for the next overnighter. No shade was available in the marina near the boat. All gas blending was done in the open under the sun. Deco gas bottles were immersed in a bucket of water, but doubles and storage cylinders had no protection from the summer heat.

To compensate for the expansion of gases being cascaded, John Yurga and Tom Surowiec - mix masters for the trip - applied what Surowiec called "Kentucky windage." This is an estimated adjustment for temperature that is based upon the mixer's judgment and experience. They overfilled the tanks with helium by a couple of hundred psi, so when the gas cooled off it would settle out to the desired component pressure. Granted that this has the flavor of ascertaining wind direction and speed by holding a wetted finger in the air, but sometimes it's the best gauge you've got.

If the storage cylinders are left in the sun, let the recipient tanks warm in the sun as well. This will equilibrate the temperature.

square one and wasting gas, or you're forced to make compromises that could have been avoided.

The good news is that once you learn the recipe and know the ingredients in exact psi, the process is shockingly repeatable.

In summary: you wouldn't want your dive shop professional to sell you a gas blend that was close but not quite right, so don't expect less of yourself.

Accuracy Counts

To achieve the fractions of constituent gases that are desired in a final mix, the introduction of precise component pressures of each gas is required. For instance, if you wanted to home-brew nitrox-39 with a final pressure of 4,500 psi, you would first have to base the tank with 873 psi of oxygen. You may ask, why do blends require an oxygen pressure measured not just in the tens, but in the ones? The reason is that the calculations are based on mathematical formulas which yield ideal blends.

"What are a few psi among friends?" you might pose. Isn't 870 psi or 880 psi close enough? Will such a slight error materially affect the CNS clock or the decompression profile? Realistically, the answer is no.

Am I being picayune by insisting on near decimal accuracy? As a rejoinder I can only state that a cavalier attitude toward mixing gas is a sentiment that is unhealthy and likely to get you into deep trouble.

By itself one minor inaccuracy may not count for much, especially when compared to other factors in which the inherent margin for error is greater: gas compressibility, pressure gauge readability, oxygen analyzer accuracy, temperature, and humidity, to name a few. If you don't strive for precision, the cumulative effect of multiple inaccuracies that are introduced at every step of the process may be considerable.

The final blend might bear no resemblance to the blend you intended to make. You're back to

The Nitrox Home-brew

The gas most often blended at home is nitrox. If you remember from several sections back, all you need to make home-brew is a storage cylinder of oxygen and a transfill hose. Cascade the required amount of oxygen into the recipient tank, top off with air, and analyze.

It doesn't take a consulting detective to deduce which clue is missing in this case. A Sherlock Holmes aficionado would exclaim that it's as singular as the dog that didn't bark in the night. Where do you get the air?

This is a touchy subject.

Oxygen and a transfill hose cost a few dollars. A compressor costs a few thousand. Of course you can always rent. For expedition purposes this may be the best way to go. Or a group of technical divers might pool their resources and purchase a compressor on their own.

Usually, however, the oxygen is cascaded at home and the air is obtained from a dive shop. This is where it gets touchy.

Shop Around

Some recreational dive shop owners have complained that technical divers are topping off tanks surreptitiously, without revealing to the shop owner that the tanks have been partially filled with oxygen. Not all dive professionals are partial to nitrox. Their apprehensions are twofold: that oxygen might flow up the fill whip and contaminate the air bank, and that they might be held liable should an accident occur to the diver.

The first concern is almost baseless. Gas will flow only from a level of high pressure to a level of low pressure. Even if the pumping facility did not employ check valves in the tube lines, gas from a scuba tank can no more flow into a high-pressure air bank than water can flow up a cataract.

Theoretically, it's possible for gas to backflow into the fill whip, but only if the shop owner or filling technician is careless. After a scuba tank is filled, the routine sequence of events is to first close the tank valve, then close the bank valve. If this procedure is reversed and the bank valve is closed first, gas from the tanks could possibly mingle with the gas in the fill whip because the pressure in both is the same. This could happen only in the few seconds before the tank valve is closed. If the technician immediately connected the fill whip to another tank without blowing out the whip, the residue in the hose would be forced into that tank. The amount of additional oxygen that would enter the recipient tank is negligible; it could be counted in parts per million.

But suppose two or more tanks are being filled at the same time? Here we can definitely have a problem. In many filling station set-ups, all recipient tanks are opened at once and then filled from a bank or compressor. When the tanks are open they will equalize. If the pressure of oxygen in one is higher than the pressure of air in another, the oxygen will transfill.

In this scenario, the resulting nitrox blend will have a lower percentage of oxygen than that intended, and some other customer will leave the shop with air that has been enriched without his knowledge. Both end-users are at risk: the former from DCI and the latter from oxygen toxicity.

As technical diving gains greater acceptance the knowledge gap will close. Shop owners and home-brewers will become better educated and will coordinate their efforts. Eventually, even the most nonprogressive dive shops might expand their operations to take advantage of the business opportunities presented by technical diving.

You can help this maturation process by acquainting shop owners with the facts. Any reasonable person can follow the logic I just presented. Don't bother trying to edify people who have unalterable mindsets: it's a frustrating exercise in futility and not worth the brouhaha. "You can't push rope," as we said in the electrical trade.

Liability, on the other hand, is an absurdity of American culture that is totally unpredictable. Almost any concern with liability can be justified. The relevance here is that the shop may be blamed for any mishap to the diver on the grounds that the shop was partly responsible because it supplied a component of the breathing gas.

There is no logical basis for this contrived and unwarranted rationalization. But then American civil courts have never been known for any display of logic, reason, or fairness. Nor are they known for placing responsibility where it belongs. They are known for extracting unjust retribution from those who have the most money or who are adequately insured.

My advice is to patronize shops whose owners are broad-minded and living in the twenty-first century. Don't hide the fact that you're making home-brew. Seek approval. Then they won't get suspicious when you demonstrate anxiety over obtaining a precise final pressure. And you won't have to worry about not getting the proper blend.

A word of advice: if you patronize a shop that already sells nitrox, buy your gas from the shop. You can't expect a shop to be receptive to the idea of giving away a service that is part of its business.

Air Grades

Another drawback to making home-brews is that unless the shop you patronize uses an oil-free compressor or has adequate filtration, your oxygen-cleaned tanks will get contaminated every time you top them off. Your regulators will be contaminated in turn.

This means re-enacting the laborious process of cleaning your equipment for oxygen service after every dive. I can guarantee that if your tanks spark off a flashover while being filled in a shop, the owner will be justifiably upset. This is not the way to earn his respect for your technical expertise. Cutting corners can cause more trouble than it's worth.

What to do? The simplest solution is to get air from a shop whose grade is pure enough to prevent contamination. Normal dive shop air is designated by the CGA as Grade E. The purity required for oxygen service is Grade J. A shop that dispenses Grade J air should display a certificate that affirms that its air has been tested by an industrial testing agency and found to meet the Grade J standard of purity.

Those who own oil-free compressors would like to believe that no compressor that is lubricated with petroleum-based antifriction products can deliver clean air. This belief is bogus.

Insufficient filtration and dirty filters leave contaminants and oily residues in the air. Any compressor can supply clean air if it has enough filtration and if the filters are changed regularly. Ask your shop owner what grade air he supplies. If he doesn't know or if he's not conversant with air grades, patronize another shop.

Alternatively, you can buy your own filtration system. This is not a minor expense, but if you're taking the route of purchasing your own compressor it will ensure that the air you pump into your tanks is clean. It's also possible that a dive shop owner might permit you to add your own filtration to his compressor line whenever you top off home-brews.

A dive shop is a business. If you appeal to the owner's commercial interests you might obtain the cooperation you need in a way that is mutually beneficial.

Heliair

Heliair is often called the "poor person's mix." This is because of all the helium mixtures that are blended for technical diving purposes, heliair is the least expensive to make and requires the smallest amount of preparation and effort to mix. It also has definite limitations.

To paraphrase a statement made a few sections back, all you need to make home-brewed heliair is a storage cylinder of helium and a transfill hose. Cascade the required amount of helium into the recipient tank, top off with air, and analyze for oxygen. Anyone who masters the mixing of nitrox can mix heliair as well.

Furthermore, tanks in which heliair is blended do not need to be cleaned for oxygen service. This is because pure oxygen is never introduced into the tank, and the partial pressure of oxygen in the final mix is less than that in air. Helium is an inert gas that is not affected by contaminants and oily residues any more than air.

This also means that tanks can be topped off at any reputable dive shop without any concern about the need for extra filtration. Heliair puts mixed-gas diving at everyone's fingertips.

Another advantage is that heliair is infinitely "retoppable." That is, you don't have to dump the leftover mix after every dive and make a new blend from scratch. Since the proportions of oxygen and nitrogen in air are fixed and immutable, you can calculate how much helium needs to be added for the next blend.

Steve Gatto has a trick for making accurate heliair blends. At home he cascades slightly more helium into his tanks than is required for the desired final mix. When he gets to the shop to top off the tanks with air, he releases the excess helium through his own precision pressure gauge until the exact partial pressure is attained. The interval between home and shop allows the helium to settle out from any overpressurization due to the heat of expansion. Furthermore, by standing his tanks next to the shop's storage cylinders - whether they be outside in the sun or in the cool of the basement - they achieve equilibrium with respect to temperature. Thus he averts inaccuracies that can be induced by temperature differentials between gases.

A good rule of thumb for making any kind of blend is: don't pick a critical mix. Leave room for error so if the resultant blend is slightly off you can still breathe the gas. Do this by shooting for a partial pressure of oxygen that lies in the lower end of the acceptable safe limit with respect to oxygen toxicity - another reason for choosing 1.4 ATA of oxygen over 1.6 ATA.

The downside of heliair is that the obtainable blends are restricted to certain specified percentages of component gases. Although heliair contains oxygen, helium, and nitrogen - the same as trimix - the proportions of oxygen and nitrogen cannot be modified with respect to each other.

In essence, by counting the percentage of oxygen in whole numbers, there are less than a dozen blends of heliair that have any practical value. In each of these blends the proportions of helium and nitrogen are immutably established by the constraints of the blending process.

For example, if you cascade 1,000 psi of helium into a 3,500 psi tank and then top off with air, the resultant blend contains 15% oxygen, 28% helium, and 57% nitrogen. The variable in the equation is the helium. The leftover space is filled with air. If you wanted to nit-pick, you could claim that intermediate blends could be obtained by carrying the percentages of oxygen to decimal

Wings to go. Wings Stocks owns and operates a technical dive shop in California called Ocean Odyssey. He takes his show on the road in the "wingsmobile": a traveling mixing station complete with booster pump and compressor. Here, after driving across the country, he's mixing gas for a trip to the *Andrea Doria*.

places, but this is mere interpolation and doesn't yield blends that are significantly different with respect to the partial pressures of oxygen or equivalent narcotic depths.

The blend in the above example is written heliair-15/28, in which the percentage of nitrogen is assumed, much in the way that trimix-15/28 assumes the percentage of nitrogen as the balance. Note that heliair-15/28 is the same blend as trimix-15/28. Chemically they are identical.

Home-brewed Trimix

Once you feel comfortable making home-brewed nitrox and heliair, you're only one step away from making home-brewed trimix. Blending trimix is a two-step process. All you need is a storage cylinder of oxygen, a storage cylinder of helium, and a transfill hose with two adapters.

First you cascade the oxygen into the recipient tank to the designated pressure. On top of that you cascade the required fractional component of helium. Then top off with air and analyze for oxygen.

I don't need to repeat myself about compressors, filters, and dive shop owners. But we're back to using equipment that must be cleaned for oxygen service. Why not cascade the helium first, you may ask, then cascade the oxygen on top of it. This procedure would minimize the high concentration of oxygen in the recipient tank. There's some truth to this scenario - a shallow truth. A good truth.

The fact is that for dives to depths of more than about 400 fsw, very little additional oxygen is needed - in the range of a couple of hundred psi - so that cascading oxygen on top of the helium is doable. For shallower dives more oxygen may be required, but if the amount of helium is small enough then the oxygen can be introduced in the second step.

For dives to depths deeper than about 500 fsw, the required fractional component of helium is so high that oxygen can be cascaded on top of it only from a full or nearly full storage cylinder. If you wanted to be tricky - and if you were skilled enough at blending by fractional components - you could cascade only part of the required helium into the recipient tank, cascade all the required oxygen, then add the remainder of the helium. I should warn you, however, that it is practically impossible to achieve any degree of accuracy by means of alternating gas induction.

In all these scenarios, as long as the percentage of oxygen in the recipient tank never exceeds 40%, this procedure would obviate the need to have the tanks cleaned for oxygen service. But keep in mind that the flashover risk in these cases is greater due to the higher pressure of oxygen going through the valves. Decant the oxygen at the pace of a crippled snail.

Trimix has one distinct advantage over heliair: the ratio of constituent gases is infinitely variable. With trimix you can optimize the partial pressure of oxygen and the equivalent narcotic depth for the maximum depth of the dive - instead of having to accept half measures that result from the unalterable ratio of gases in heliair.

Brewer's Yeast

Just as beer requires yeast to ferment the brew, accurate gas blending demands the leavening of experience. I know it looks simple on paper, but in practice you'll find that a certain amount of finesse is required to achieve the proper mix. The best teacher is a written record of your mistakes.

You can't become an expert gas blender just from perusing a book - even this book - any more than you can race in the Indy 500 the day after reading a driving manual. So don't get frustrated if your mixes come out wrong at first. It's all in the wrist, as they say.

If you can't find anyone to show you the ropes, try making practice blends in a pony bottle. Mixing gas in a 30-cfg bottle is the same as mixing gas in a 120-cfg tank - the quantity of gas is different but not the pressures of the component gases. The cost of exotic gas you'll waste in the process will be cheap compared to the value of the experi-ence you will gain. Once you master small scale blending you'll be ready for the big leagues.

Get with the Program

By now it must have crossed your mind that I haven't added the most essential ingredient to the recipe for mixing gas: the fractional proportions of constituent gases. If you've gotten this far in technical diving and are breathing helium blends, then you probably already have that data. They are embedded in your home computer decompression program - the one you use to cut your decompression schedules.

If you haven't yet purchased such a program, but are gleaning ideas from this handbook on how best to proceed into the deep, then I suggest that before you invest in the software you read the promotional literature on the various programs available (about a dozen exist but they come and go) and choose one with user friendly gas blending capability. Or poll some technical divers and technical dive shop professionals for advice.

Alternatively, an inexpensive software program called Mix Master can provide the partial pressures for any blend of nitrox, heliair, trimix, and heliox. Just plug in the data that you want invariable and the program does the rest.

Tables exist in widespread circulation that furnish the partial fill pressures for mixing nitrox and heliair and for certain blends of trimix. The problem with these tables is that they are derived from linear equations based on ideal gas laws. These tables do not take into account differential gas compressibility. This means that the fractional component yields may come up short.

You can waste a lot of gas while forming fudge factors to create tables that really work. Remember what I said about the importance of precision. If you have to overestimate each fractional component so the gases settle out to the correct fill pressure in the final blend, you've added another quantum of uncertainty to the process.

Sure you'll catch the error by analyzing for oxygen, but that doesn't mean that you can readjust the blend. The only workable correction method is to dump some of the final blend and top off again with air. If that doesn't work you have to blow down the tank and start over.

The software programs mentioned above take differential gas compressibility into account. Or at least they should - check first before purchasing. They do this by introducing a variable component into the computation equation. If you compare every program on the market you'll notice that each one yields a slightly different result.

You might expect uniformity from calculations based upon a sound scientific principle. But do not be disturbed by minor dissimilarities between programs. More inaccuracy is introduced in the cascading process (due to gas expansion, gauge readings, ambient temperature, and so on) than by software eccentricities.

Differential Gas Compressibility

Take a batch of wooden blocks that children use in play. If each cube is two inches square and you stack five of them in a tier, the height to the crown of the highest block is ten inches. This linear addition is analogous to ideal gas law because the blocks are solid matter.

But in reality gas is compressible. Think of gas as blocks of foam rubber. If you stack five two-inch rubber cubes in a tier, the weight of the upper cubes presses down on the lower cubes. Only the top cube retains its original dimensions. The cubes beneath it are flattened or compressed so that the height to the crown of the highest block is somewhat less than ten inches.

To make this analogy more comparable to gas compressibility we must press the foam rubber cubes between our hands instead of simply stacking them atop each other. This is because in a mixed gas *all* the constituent gases are compressed. Furthermore, our analogous rubber cubes must be made of different density foams - some more compact than others - because every gas has its own compressibility factor.

Chemists overcome this problem by mixing gases by weight. But this can be done only in laboratories or commercial blending facilities that have molecular scales or sophisticated measuring instruments.

You can't detect gas compression during the mixing process while adding fractional components. It becomes apparent only after the tanks are full and you analyze for oxygen. Then you discover that the percentage of oxygen in the tank is somewhat more or less than what was intended. By extrapolation, the percentage of inert gases must also be off the mark.

The fraction of inert gas can deviate by several percentage points without significantly altering the decompression profile or equivalent narcotic depth. But the percentage of oxygen needs to be more precise due to the risk of oxygen toxicity. A 1% or 2% overage could be fatal at depth.

To make matters worse we must now add another variable that goes beyond the foam block analogy: the order in which the gases are cascaded into the tank. Let me explain first by giving an example.

When 200 psi of oxygen is introduced into an empty tank, it will result in a lower fractional component of oxygen in the final blend than when 200 psi of oxygen is introduced into a tank that is already holding 1,000 psi of helium. This condition occurs because in the first scenario the oxygen is somewhat compressed by the helium, thus yielding a lower fraction of oxygen in the final blend. In the second scenario you are actually adding more then 200 psi of oxygen because the oxygen is being compressed as you introduce it on top of the pressure of the helium, thus yielding a higher fractionl component of oxygen in the final blend.

In truth, the whole is more than the sum of its parts because the parts have been compressed during the process of blending. It's as if you pressed down harder on the tier of rubber cubes whenever you added another cube, or whenever you added more foam to an already existing cube. Another way of looking at it is like this: if you could magically merge 200 psi of oxygen with 1,000 psi of helium (both pressures based upon a base pressure of zero) the result would gauge out at less than 1,200 psi.

You can't watch this phenomenon occur because the pressure gauge is part of the system. The gauge cannot show the fractional components of individual gases. It registers only total pressure, and that on a linear scale.

As if this isn't enough, gas compressibility is also dependent upon the final pressure of the mix. This means that the fractional components of individual gases will be different in a tank that is filled to 2,400 psi from a tank that is filled to 3,500 psi.

Add temperature gradients and Surowiec's Kentucky windage, and you can understand why blending gas is not as easy as mixing one part gin to one part vermouth to make the perfect martini. Blenders beware.

The Multiple Cylinder Cascade System

So far I have described how to cascade gas from only a single storage cylinder. But what do you do when the pressure in that cylinder of very expensive helium drops below the component pressure you need to introduce into your tanks?

The obvious solution is to chalk "MT" on the shoulder of the cylinder and exchange it for a full one. An obvious solution but a wasteful one, because you don't get credit from the commercial gas supplier for leftover helium that is returned. Your loss is their gain.

But by setting up a multiple cylinder cascade system you can eke out nearly every last psi of helium that you paid for. I'll explain by way of example using convenient and easily divisible numbers. Let's say the storage cylinder starts out at 2,400 psi and that after a couple of fills it's down to 1,000 psi. (I'm going to talk in partial pressures only, not cubic feet of gas.)

Connect the transfill hose to an empty tank (or set of doubles). Let's say you want 1,000 psi of helium in the recipient tank. You can't cascade that much gas because the storage cylinder and the recipient tank equalize at 500 psi. So you connect the recipient tank to a full cylinder of helium and cascade the rest of what you need. If the storage cylinder and the recipient tank were the same size (that is, held the same number of cubic feet of gas when full - which we will assume in order to simplify this example) the pressure in the second storage cylinder would be reduced to 1,900 psi and the recipient tank would be ready for topping with air.

This leaves us with two storage cylinders of helium: one with 500 psi and the other with 1,900 psi. For the next fill we connect the empty recipient tank to the storage cylinder with the lower pressure and equalize them. Now we've got 250 psi in each. Cascade the rest of the helium (750 psi) from the second storage cylinder. The recipient tank is now ready for topping off, and the second storage cylinder is down to 1,150 psi (1,900 psi minus 750 psi).

If we repeat the process a third time, the first storage cylinder will equalize with the recipient tank at 125 psi (half of 250 psi) and the second storage cylinder will bring the pressure in the recipient tank up to 637.5 psi (midway between 125 psi and 1,150 psi) at which point the second storage cylinder is also equalized. A third storage cylinder must now be brought on line in order to make up the difference to 1,000 psi in the recipient tank.

On yet another fill the pressure in the first storage cylinder drops to 62.5 psi. You can go half again, but at this point it's economically acceptable to exchange the cylinder for a full one and move it from the beginning of the cascade system to the end. Thus three cylinders in rotation is the minimum number necessary to utilize such a system - assuming that exchanging cylinders can be done quickly and conveniently.

It is important to note that the example above works when no more than about 1,000 psi of helium is required in the recipient tank. However, as receiving pressure rises (for mixes that contain more helium) more bottles must be added to the cascade system in order to draw the lowest bottle to below 100 psi. For example, a recipient tank that requires 1,400 psi of helium will require four bottles in the cascade system in order to eke out that last 100 psi.

The multiple cylinder cascade system is just as effective for cascading oxygen and argon as it is for cascading helium. Very quickly you could find your garage filling up with banks of bottled gas.

The Booster Pump

The Haskel gas booster is a simple but effective mechanical device that elevates the pressure of a gas while transferring it between cylinders. In plain language, the booster pump sucks the gas out of one tank and spits it into another, at a higher pressure. It does this astonishing feat through pneumatic advantage.

Randy Bohrer of Underwater Applications rebuilds and sells Haskel gas boosters. He described them succinctly in his catalogue: "These units are capable of boosting gases from a few hundred psi to several thousand psi. The power source is low pressure, high volume air or gas which operates a large piston valve assembly. This drive piston is directly connected to dry running, hydrocarbon free pumping pistons. The area of the drive piston exceeds the area of the gas piston by various ratios (depending on the model) which allows the unit to compress gas to thousands of psi with only 100 psi of drive gas."

Note that this remarkable performance is achieved without electricity, without batteries, without chemical fuels, without windmills.

For dive shop applications a booster can be driven by high-pressure air that is diverted to the unit from the storage bank. A regulator reduces the air to the required drive pressure.

For remote land-based operations a booster can be driven by an air compressor. This was how we boosted deco gases and helium mixes on the 1994 *Lusitania* trip in a secluded corner of Ireland. On a boat over the *Andrea Doria* I saw Bohrer drive a booster with compressed air from an ordinary scuba tank.

The Haskel booster is truly portable. It is compact and light enough to be lifted and carried by a single person. The cost of a new unit is about half the price of a mini-car. Boosters can be rented at reasonable rates for dive trips and expeditions.

If this sound too much like boosting by your bootstraps, it very nearly is.

The major constraint of the booster is the volume of air needed to drive it. While only a minimal amount of drive air is required to boost gas from 2,500 psi to 3,000 psi, boosting gas from 2,000 psi to 3,000 psi requires twice as many cubic feet of drive air per cubic foot of gas boosted. Boosting gas from 1,000 psi to 3,000 psi requires four times as much drive air per cubic foot of gas boosted as compared to the first example.

Simon Tapson and Nick Hope are mixing gas for a dive to the *Lusitania*. The booster in the homemade wooden frame was driven by the low-pressure compressor on the trailer, which also drove a second booster whose manufactured aluminum frame is visible over Hope's shoulder.

Boosters are capable of consuming as much as 60 cubic feet of air per minute. For that reason it should be driven by a high-volume low-pressure compressor.

A booster can be driven by any bottled gas on hand, not just air.

Gassing up for the Lusitania

The 1994 *Lusitania* project was the grandest and most widely publicized wreck diving trip of the decade. For conduct and complexity it deserves true expedition status. The project was executed successfully not by accident but by very definite design. This was due to two very important elements (in addition to oxygen and helium): meticulous planning and methodical organization.

The core group that initiated the venture consisted of eight U.K. divers who called themselves the Starfish Enterprise. The undisputed leader of the Starfish Enterprise was Polly Tapson. Each of the participants was delegated a job or specialty for which he or she was responsible. Regular meetings were held at which everyone reported on the progress of their assigned tasks. (The four Americans were excused from attending.) Tapson chaired these meetings and coordinated activities. The trip was a year in the planning.

There were no technical dive shops anywhere near the south coast of Ireland. This meant that we had to handle all our own arrangements for procuring gas and buying, borrowing, or renting equipment. (We is figurative - the "Brits" managed the preparation work; the Americans could not lend assistance until we arrived in London the week prior to departure for Ireland.)

The wreck lay at about 300 fsw (deeper during high tide). For that depth "we" ordered pre-blended trimix-14/43. This was delivered by lorry to the folk house that served as our base of operations for the fortnight. The owner of the folk house had carpenters build a temporary wooden addition in which the gas storage cylinders were stored and secured. The gas cylinders filled two wooden pallets, sixteen cylinders per pallet (called a "quad" because the cylinders stood in four rows of four.)

The decompression profile required three deco gases: oxygen and two blends of nitrox. We had to blend our own nitrox every day for the next day's dive. We also had to transfill trimix daily from the storage cylinders to our scuba tanks. For these jobs we used two Haskel boosters that were driven by a low-pressure, high-volume air compressor that was towed to the site on a trailer.

With all of us working to the fullest capacity, it took every spare minute of every day to dive and boost the gas for the next day's dive. We rose early each morning, made last minute preparations and analyzed gas, rode the boat to the wreck site, conducted dives which had a two-hour run time, returned to shore, and toiled far into the night boosting gas for the next day's dive. It was exhausting labor.

The work load increased after a week when our pre-mix ran out. We had more helium and oxygen delivered. But we couldn't just boost the trimix as we had been doing - we had to blend our own.

If all this sounds like a lot of work, that's because it was. Expedition diving is not Club Med.

Loading the boat for a dive to the *Lusitania*. The red inflatable carried the dec station and was used as a chase boat.

Oxygen Extractors and Concentrators

It may sound like science fiction but it's for real. Once upon a time an inventor developed a method for extracting nitrogen from air through a gas permeable membrane. The nitrogen was stored in cylinders while the waste product, oxygen, was exhausted into the atmosphere. Then someone realized that oxygen might be a salable commodity too.

Thus was born the oxygen generator: not an electrical device that is powered by oxygen, but a high-tech mechanism for generating oxygen by extracting it from air and concentrating it in cylinders. Two varieties exist.

The DNAx is primarily a "continuous nitrox mixer" designed for commercial applications on the scale of dive shops and live aboard dive boats. The unit delivers "denitrogenated air" by means of differential permeability technology. Dick Rutkowski described it best:

"This concept passes compressed air through filters with condensate traps to remove moisture, oil, and particulate carryover. Membranes employ the principle of selective permeation of separate gases. Each gas has a characteristic permeation rate that is a function of its ability to dissolve and diffuse through a membrane. This characteristic rate allows 'fast' gases to be separated from 'slow' gases."

The membrane air separator consists of thou-sands of hollow fibers that "provide a very large membrane surface for air separation." There are no moving parts in the separator. However, a compressor is required to pull air through the bundle of fibers. Only 140 psi is required to make the membrane work. The membrane is not capable of handling high-pressure air, so the gas must be compressed after leaving the membrane. Thus the system works in two stages: one to separate the oxygen, another to compress the oxygen into storage cylinders.

A blending station can be established with supplementary hardware. The end product is nitrox whose oxygen content is less than 40%. This means that neither the tanks, the compressor, the fittings, the hoses, nor any affiliated transfer equipment need to be cleaned for oxygen service. Since the nitrox is made without resorting to bottled oxygen, this eliminates the handling of heavy storage cylinders.

Promotional literature stresses Grade E filtration, trimix capability with the addition of customized system assemblies, and remote site use with a gasoline powered compressor. The quantity of oxygen generation is not particularly fast - it may take all day to fill a single storage cylinder - but the unit can run unattended.

The InstaGas Oxygen Concentrator utilizes molecular sieve adsorbents and "the method of pressure swing adsorption (PSA) to preferentially adsorb undesired gases in a compressed air stream." The advertising brochure explains that "compressed air is applied to an adsorber column . . . which is filled with molecular sieve. As the compressed air flows through the column, the undesired gases are adsorbed by the molecular sieve, while the desired gas is allowed to pass through the column, where it is extracted as concentrated product gas."

InstaGas can concentrate oxygen that is dry and up to 95% pure, and at a flow rate of five liters per minute. The system comes complete with a choice of booster pumps that can amplify oxygen pressure as high as 3,000 psi. Because of the high percentage of oxygen it is recommended that the system be operated "from an oil-free compressed air source with a minimum supply pressure of 60 psi."

While the booster pump and molecular sieve are driven pneumatically, the controls are electrical, so a source of electricity is required in order to operate the system. For remote diving purposes, alternating current can be supplied by a small generator. Drive air can be pumped from a portable compressor.

The modular design enables InstaGas to be customized for mixed-gas applications.

The cost is about the same as that of a small automobile.

Both DNAx and InstaGas can supply 100% nitrogen, in case you have a need for it.

Emergency In-water Oxygen Recompression

Perhaps nothing in technical diving is so highly controversial as the procedure for conducting omitted decompression in the absence of a chamber. This method was pioneered in Australia where commonly dived reefs are many miles from shore, and where the nearest shore was nothing more than uninhabited wasteland. The distance to the closest metropolitan area large enough to support a hospital with a chamber facility might be a thousand miles or more - and this without means of rapid transportation or emergency medical evacuation.

Here the meaning of the word "remote" is nonpareil.

A diver who showed symptoms of DCI after a dive had only two possible alternatives: suffer the consequences of a malady that went untreated (paralysis or death) or go back into the water and breathe oxygen under pressure.

Physicians and those who have gone unafflicted have historically ridiculed any attempt on the part of a patient to save himself by whatever means available. Their attitude is based upon the premise that it is unsafe to conduct recompres-

sion in open water where conditions and medical care are not under strict scientific control.

Those in danger of dying or spending their lives in a wheelchair chose a more realistic expedient, opting to take a chance on a possible cure against the almost certain lethality of delayed or absent treatment. The system thus devised *for a state of dire emergency* was one which relies upon the beneficial purging effect of oxygen in eliminating nitrogen.

Without a doubt the procedure is risky due to the chance of losing consciousness under water as the symptoms of DCI continue to manifest. A full-face mask or an unafflicted observer can help prevent the patient from drowning as a consequence. Positive buoyancy can add another margin of safety: if the decompressing diver passed out, his hand would relax from the downline and he would automatically rise to the surface.

Emergency oxygen recompression tables have been published. The technical diver who plans to dive in remote areas would be well advised to obtain a set, and to be prepared to carry out the procedure should the necessity arise.

Since those tables were designed primarily for non-technical divers, they may not take into account the excessive exposure to oxygen that a technical diver may already have incurred. Thus the use of such a method of recompression might include the additional risk of oxygen poisoning.

Make allowance for this by planning remote dives more conservatively. Do not push the CNS clock. Leave room for later exposure to oxygen by reducing the ppO2 of the dive to 1.4 ATA or less. Conduct accelerated decompression on blends such as nitrox-70 instead of pure oxygen, so the CNS clock does not max out during deco. Add fudge factors to your decompression profile. Hang a little longer than necessary. Don't do extraordinary work too soon after a decompression dive.

Prevention is the best medicine. Even doctors can't disagree with that.

UNDER EXTENDED PRESSURE
Portable Recompression Chambers and Decompression Habitats

It can be calculated mathematically that if a thousand diggers can excavate a canal in ten years, then one digger can excavate the same canal in ten thousand years. Logic dictates otherwise.

As one who enjoys the small scale aspect of solo diving, in which I can do as I please without having to divert my interests by working on a project whose goals are not my own, I recognize and have participated in group endeavors whose objectives were impossible to attain alone. Even as a very private person I cannot ignore the feeling of camaraderie that comes from sharing a vision with others of like sentiment, from combining the spirit of adventure with the inspiration of friendship, from experiencing the joy and satisfaction of accomplishing tasks through teamwork. It is that innate evolutionary characteristic that has vaulted mankind from such meager beginnings to a stellar force.

"Pushing" a deep submerged cave system while breathing mixed gas demands great cooperation among individuals. The location and exploration of deep water wrecks are coordinated ventures. Such projects cannot be tackled independently. Certain objectives can be achieved only by working in cooperation with others.

We have entered a new age of underwater exploration in which the collective effort reigns supreme.

People pooling their resources can share the workload and expenses. Grants may be obtained from non-profit organizations, educational institutions, or government foundations. Sponsorship can be procured from companies that are interested in advertising their products. Donations may be solicited from rich benefactors seeking tax deductions.

All these means and more are available to the creative explorer. That's how expeditions were mounted to traverse the Atlantic ocean, to explore darkest Africa, to attain the Poles, to climb remote mountain peaks, to plumb the ocean depths, and to reach the Moon and beyond. A modern explorer needs more than the will to explore, he needs the money and the enthusiasm of his backers.

With the amount and the extremes of deep dives being conducted today one cannot help but look into the possibilities of how to prepare for every contingency in safeguarding health, and of how to make the inevitable long decompressions more comfortable. Most people don't have access to recompression chambers or keep decompression habitats in their back yard. The expense is beyond the means of the average-income diver.

But when divers unite their energies in order to achieve a common goal, they can go far and very deep. Keep this thought in mind when considering the cost of the equipment described throughout this chapter.

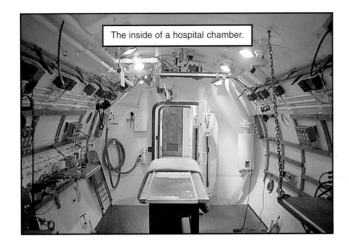

The inside of a hospital chamber.

A Chamber Made to Order

The primary difference between a *de*compression chamber and a *re*compression chamber is one of purpose.

Generally, a pressure chamber that is employed to decompress Navy or commercial divers after working under water is called a *de*compression chamber. A pressure chamber that is employed to recompress a diver stricken with a high-pressure malady in order to decompress him again, gradually, is called a *re*compression chamber.

A pressure chamber that is used for treating patients by means of immersion in therapeutic doses of high-pressure oxygen is called a hyperbaric chamber (or hyperbaric facility) or a treatment chamber.

A secondary difference among pressure chambers is location. Decompression chambers are usually found on Navy ships, oil-rig platforms, and salvage vessels. Recompression chambers and hyperbaric facilities are most often situated in hospitals.

In actual practice, a decompression chamber can be used for recompression, and a recompression chamber is often the same chamber that is used to treat patients no matter what their disorder. And a decompression chamber might be used to immediately recompress a diver whose decompression proved ineffective in preventing the bends.

The precise designation is "pressure vessel for human occupancy." In keeping with the penchant of the scientific community, this linguistic form is specifically descriptive and typically cumbersome.

All of these variations on a theme can be con-

fusing to those who aren't in the business. In common usage the distinctions are unimportant and irrelevant. "Hyperbaric chamber" is a generic term because "hyperbaric" means "high-pressure." Most of the time even the professionals abbreviate the whole gamut of nomenclatures to the simple word "chamber," leaving one to wonder if the nurse or technician who tidies up after each use or treatment is known as a "chamber maid."

The chamber above is a deck decompression chamber in use onboard a vessel during a commercial salvage operation. It doubled as living quarters for divers who remained saturated at the working depth for the duration of the job. At right is the bell which mated with the chamber and transferred divers to and from the bottom.

The Emergency Chamber

I hate to add another variation to an already overworked theme, but the chambers I want to discuss in this chapter are designed for emergency use only and are moveable.

Moveable chambers generally cost about as much as a large house or an unpretentious mansion. They are expensive to maintain and can be operated only by trained technicians. For diagnosing pressure-related ailments, a hyperbaric physician must be in attendance. The doctor determines which emergency decompression procedure the technician needs to implement.

These chambers are so huge and heavy that

they must be transported on flatbeds (either a train or tractor-trailer), then hoisted by crane aboard a fair-sized ship (not a boat). Furthermore, they can't be pressurized without a vast supply of bottled gas and/or a large capacity compressor. So you can understand why only governments and large companies can afford to own or rent them.

But there's no rule stating that emergency chambers need to have multiple person capacity or double lockout capability. Just as there are fully staffed hospitals and first aid kits, there are chamber complexes and single-person units.

The Portable Chamber

Portable chambers have been around for a long time. Because of their confining nature they are usually called "pots." People who have been forced to spend time inside them reminisce not-too-fondly about being pressure cooked in the hot sun on the deck of a boat by their cannibalistic cohorts. A single-person chamber is like a cylindrical coffin; only Ray Milland in *The Premature Burial* could be happy inside one.

Nevertheless, a person feeling the pain of DCI or facing the possibility of partial paralysis is better off fighting his quiet claustrophobia than suffering from noisy bubbles in the blood. For diving expeditions in remote locations where evacuation is a national event, the portable chamber is the only way to go.

I picked up a telescoping chamber that had outlived its usefulness as a laboratory tool when grant money materialized at a local university for a new and improved model. Like a collapsible drinking glass designed for travelers and backpackers, the aluminum chamber consists of five interlocking sections, each sealed with a gasket against the adjacent section. Packed

The self-contained Hyperlite chamber uncrates like a Jack-in-the-box and comes with its own control panel. The tank of air is for pressurization, the oxygen is for decompression treatment. (Courtesy of SOS Ltd.)

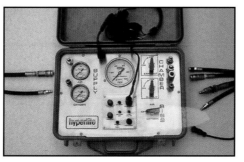

away, it looks like a tree planter. Telescoped open, the chamber is big enough for a six-footer with a crew cut who is not overthin. It has a window the size of an ashtray and tubing for pumping oxygen into a breathing mask.

Like its bigger brothers, the telescoping chamber requires either bottled air or a compressor to bring it up to pressure. Even then, it's not made to hold as many atmospheres as the commercial models. Remember, though, that the purpose of a portable chamber is to provide first aid and to stabilize a diver stricken with a pressure related injury. It is equivalent to the compress placed on a wound before a doctor can sew in stitches. Fortunately, I've never had to use my chamber.

The Inflatable Chamber

Whereas my chamber is constructed of thick-walled aluminum and is relatively heavy, a British company has put on the market a chamber that is truly innovative in design. SOS Ltd. manufactures a "Lightweight portable and flexible hyperbaric chamber" called the Hyperlite System. It comes in one-person and two-person models. The fascinating feature of the Hyperlite chamber is that it is inflatable.

According to J.S. Selby, Managing Director of SOS Ltd., the Hyperlite chamber "was not designed as a therapeutic unit, but as a transport facility to take the patient, under pressure and breathing oxygen to the nearest fully manned therapeutic chamber for treatment with full medical supervision." Think of it as a hyperbaric stretcher.

The hull of the Hyperlite chamber is made of Kevlar fiber embedded in a silicone rubber matrix that is stable under a wide range of temperatures, is highly resistant to abrasion and ultraviolet degeneration, and is extremely flexible. The seamless tube is free standing: that is, it maintains its tubular shape without pressurization.

Concave, clear acrylic endcaps double as windows and as points of penetration for gas hoses. The chamber and its control box fit into two storage cases small enough to be carried aboard a commercial airliner as personal luggage.

Set-up time is counted in minutes. After the patient is comfortably in place, the chamber can be pressurized to an equivalent depth of 75 fsw from an ordinary scuba tank, with air to spare. Oxygen can be fed to the patient through a bulkhead fitting and breathing mask, and there are electrical connectors for an intercom and medical monitoring devices.

Chocks keep the chamber from rolling. Using nylon straps with hand loops, four porters can easily pick up both chamber and patient. Lifting eyes allow for helicopter rescue.

The Hyperlite chamber is flexible in more ways than one. A diver forced out of the water with inadequate decompression can complete his decompression lying down. Someone with a slight pain in the elbow can take a short trip "down" to determine if the pain eases up. A more seriously stricken diver can be pressurized, then placed in a truck and transported overland to a larger hyperbaric facility. You might call this activity "decompression driving."

The diameter of the Hyperlite chamber was specifically restricted so it will fit *under pressure* through the hatch of any civilian or military chamber. It doesn't have to be unsealed until the pressure in the therapeutic chamber has been equalized with that of the Hyperlite.

The inflated chamber can also be flown in an unpressurized airplane cabin.

"All exhaled gases are removed from the chamber through the overboard dump, so that there should not be any appreciable build up of carbon dioxide or free oxygen inside the chamber at any time."

Now if you could only get meals and room service . . .

SOS Ltd. is no relation to the Italian company that manufactured the decom meter described in the chapter on "Submersible Air Decompression Computers."

Other portable inflatable chambers of similar design are either on the market or in the planning

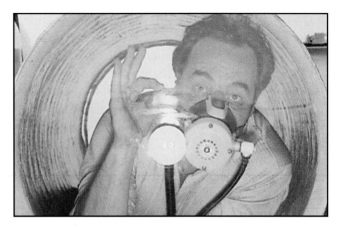

All photos on this page courtesy of SOS Ltd.

stages. One strong competitor is the Flexible Hyperbaric Chamber manufactured by the Italian firm Giunio Santi Engineering. GSE produces several models that differ from each other predominantly by their diameter. The translucent hull is made of polyester fiber. The frame is cross-hatched with polyester webbing and is inflated separately, enabling the chamber to retain its

After a stricken diver signals through the clear plastic endcap (facing page) that he is okay, a portable chamber can be carried by hand, placed in a truck for transport, or picked up by helicopter, then loaded into a full-sized chamber. The two photos on this page show a Flexible Hyperbaric Chamber (courtesy of Giunio Santi Engineering, s.r.l. and J.B. Hughes and Associates).

tubular shape when the door is open. The basic chamber features one permanent endcap and one aluminum alloy door assembly, although for greater versatility a door assembly can be fitted to both ends.

Chambers like these cost about the same as a full-sized car or mini-van.

The Decompression Habitat

Perhaps "submersible decompression chamber" is more descriptive of what the following sections discuss, but "habitat" lends an air of permanence that is bound to appeal to those who have spent many hours inside a plastic bubble whose controlled, slowly diminishing pressure change was all that prevented them from becoming soggy pretzels.

In military and commercial diving operations nearly all decompression is done inside submersible chambers that can be entered from under water. Sometimes these are used as transfer chambers: the hatch is locked so the chamber can be raised from the water and placed on the deck of a surface support vessel, where it is then mated to the hatch of a more capacious, live-in chamber into which the crew moves either to complete their decompression or to remain fully saturated for long-term projects.

This type of system is obviously beyond the resources of most individuals. However, there are other ways that habitat-type apparatus can be built, established, and utilized. In fact, habitats have been made out of a variety of secondhand containers: an old locomotive boiler, an abandoned railroad tank car, a discarded industrial cauldron, a converted milk tank, an obsolete sugar crystallizer, a thrown away pressure vessel, even a used cement mixer. Let your imagination be your guide.

Rub a Dub Dub . . .

Perhaps the simplest device I've seen is an upside down plastic tub conveniently placed inside the entrance of a cave. It was jammed against the ceiling by the air with which it was filled. There was enough room for two divers to get their heads out of the water at the ten-foot stop after a long penetration.

The air tasted slightly stale, but it required little effort to "recharge" the receptacle with a stage bottle of fresh air. You can wile away the time by debriefing the dive, having a snack, or just enjoying the luxury of not having a mouthpiece distending your jaw.

I am told that, because of their length, inverted pig troughs make excellent decompression vats - but be sure to clean them out thoroughly before installation. You wouldn't want to get a sty in your eye.

The Air Pocket

Obviously, such a device won't work in wreck diving situations unless part of the wreck lies in shallow water, or is run up on shore. But a plastic tub can be used for other purposes.

At one time I made so many dives in the same compartment of the USS *San Diego*, a World War One armored cruiser sitting turtle in 110 fsw, that my exhaled air created a shallow air pocket overhead. Visibility in the pocket was unmarred by silt.

The bow of the *Chester A. Congdon*, which lies upright in Lake Superior off primitive Isle Royale, has trapped the exhaled air of many visiting divers, and has created a large air bell where people can float and talk in a voice that sounds as tinny as that coming through the diaphragm of a cheap, off-brand telephone.

These examples demonstrate that a team of divers exploring an intact wreck could install a tub not for decompression purposes but for communication advantages. Add a wire to the surface and a diver could ask for tools, could tell the next team what to look for and where, or could exchange information with his buddy in a manner more precise than hand signals.

Wakulla Spring project. Shown above is the tubular aluminum frame being assembled in the workshop. Shown below is the habitat in position under water. (Both photos courtesy of Cis-Lunar: upper photo by Bill Stone, lower photo by Pete Scoones.)

The Wakulla Spring Habitat

As a prime example of what can be done concerning decompression habitats, let's take another look at the Wakulla Spring project. Bill Stone designed what can best be described as an underwater tent large enough to house six divers in comfort.

Across a tubular aluminum frame ten feet in diameter was stretched a pneumatic shell "fabricated from ballistics nylon that had been pressure-laminated with neoprene." The hemisphere had a floor grate with an opening wide enough to admit a diver fully dressed.

A two-ton anchor block provided bottom ballast. Attached to the underside of the dome

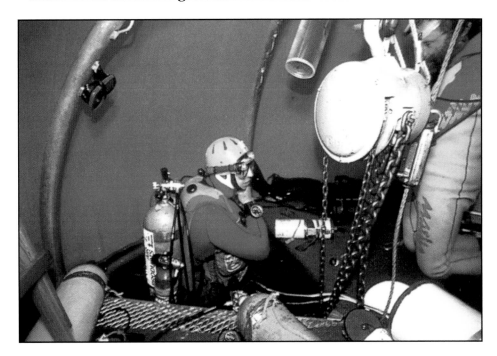

Wakulla Spring project. Shown above is an interior view of the habitat with a diver standing in the hatch. Notice the block and tackle used to raise and lower the habitat. Below right is a split-frame view with the lens bisecting the interface, depicting both the interior of the habitat and the exterior beneath it. Below left is a view of the habitat in the distance. Notice the divers and DPV's on the ledge before the drop-off into deep water where the habitat is moored. (All photos courtesy of Cis-Lunar: upper photo by Bill Stone, lower right by Wes Skiles, lower left by Pete Scoones.)

was a seven-ton ballast pad that was assembled under water by securing lead-filled pipes to a carrier.

The beauty of the system was that the dome could be raised and lowered as needed. A chain hoist inside the living quarters enabled decompressing divers to control the ascent to succeeding stages. Divers returning from exploration dives could decompress in comfort, shuck off their drysuits partway, eat a sandwich, and breathe oxygen through masks.

If all this sounds too simplistic, that's because my description must necessarily be so. What I have sloughed off in a few concise paragraphs required an incredible amount of time, planning, coordination, and the will and hard work of many experienced divers. For full details, read Bill Stone's book on the project.

Wakulla Spring project. The habitat is moored in deep water beyond the ledge on which the diver is kneeling. The ledge was used as a staging area. (Courtesy of Cis-Lunar, photo by Pete Scoones.)

Life in a Liftbag

Jim King fashioned a scaled-down version of the Wakulla Spring habitat for deep cave exploration into Dipolder. He inflated a 5,000-pound Carter liftbag which was chained to heavy weights on the bottom. Then he installed a platform which had a hole in the middle. Seats on opposite sides of the interior allowed two divers to sit out their decompression while breathing oxygen through long hoses that were attached to external cylinders.

The four-foot-high "habibag" may seem cramped, but it's better than hanging in the water for hours on end. A one-person model can be made from a smaller liftbag.

Can a similar device be adapted for wreck diving applications? I guess that depends on the wreck: its location, depth, and exposure to variable conditions (both topside and on the bottom). And to a large part on the resources and ingenuity of the team engaging the project.

According to an anonymous wise person: "The only impossibility is impossibility itself."

REBREATHERS
Exotic Breathing Apparatus

The self-contained underwater breathing apparatus that was originally designed as an everyday system for delivering ordinary air on demand has evolved into a high-tech equipment industry. High-pressure cylinders, sophisticated regulators, and state-of-the-art instrumentation can handle depths and flow rates craved by modern-day explorers whose ambition to dive deeper and stay longer has inspired manufacturers to develop the desired hardware.

Perhaps mankind's ultimate goal is to achieve complete equilibrium with the underwater world, to be able to stay indefinitely - indeed, even to live - in the marine environment without surface support. Mammals such as whales and porpoises have returned to the sea through a quirk in evolution, but they must surface periodically in order to breathe air - a compromise that makes them only semi-aquatic.

With future science and technology it may become possible to adapt people permanently to an ocean existence by surgically implanting gills, or by genetic engineering. Either path, however, is a one-way trip to the ocean floor with no way back.

Dreams and Aspirations

Beginning in the 1960's, scientists worked in earnest to develop an artificial gill that would enable man and woman to remain underwater for extended lengths of time without requiring bottled supplies of oxygen. Several research programs promised achievable solutions.

One such outcome was a "sponge" that was capable of extracting oxygen from water by breaking down the hydrogen-oxygen molecule. Few details were given. Presumably, the water molecule was cracked chemically. The impracticability of the sponge was its mass: in order to extract enough oxygen to support a person's metabolic processes the sponge had to be twice the person's size.

Another reported breakthrough was the molecular sieve. This device transferred oxygen through a filament by means of the pressure gradient created by the press of water when the diver inhaled. As with the sponge, the drawback was the monstrous size needed to achieve the supply demand.

The Russians are reported to have developed an artificial gill that regenerated air by pumping it through a gas-permeable polymer film. Although the gill was only the size of a shoe box, the auxiliary systems were unmanageable for anything less than submarine or habitat applications.

A similar concept (in theory only) involved a silicone membrane that could extract oxygen from sea water and pass it inward at the same time that it passed out carbon dioxide. Underwater Gortex, or too much wishful thinking?

One artificial gill that has actually been patented and is undergoing further research and development is called the hemosponge. This device combines organic hemoglobin with a polymer "to form a polyurethane substance similar to that in seat cushions. The sponge retains some of the biological properties, particularly its ability to bind oxygen."

Oxygen is extracted from sea water that is passed through the hemosponge, then released by means of a weak electrical charge. Biochemists calculated optimistically that the physical proportions of a hemosponge and attendant hardware sufficient to supply the respiratory needs of an individual could be carried on a diver's back.

Electrolysis is a mechanical method of splitting water into its component elements. Nuclear submarines can generate oxygen for their crews as long as their atomic fuel holds out. But the machinery is massive, requires vast amounts of electricity to operate, and cannot be housed in a portable container with sufficient capacity to meet a diver's requirement for oxygen.

In addition to their bulk, the sponge, the sieve, the film, the membrane, and electrolytic decomposition devices have one serious limitation: they are designed to supply only oxygen and make no provision for inert gas filler. Unless the hydrogen byproduct of water can somehow be utilized, the potential for contracting oxygen toxicity restricts the useful depth range of all oxygen extractors to 25 fsw.

Furthermore, they make no allowance for controlling the oxygen supply with respect to the diver's ever-changing metabolic consumption.

The One-Atmosphere Suit

One way to achieve extended bottom time is to dive in a pressure suit which, unlike its aeronautical counterpart, maintains a constant pressure by keeping out the crushing force of water. One-atmosphere suits have been around for centuries.

The earliest recorded model was one invented by John Lethbridge in 1715. It was constructed of wooden staves and metal bands much like a barrel; it even looked like a barrel. Leather gauntlets protruded from the bottom in the same fashion as the gauntlets in a radioactive glove box. A small glass window permitted the diver to see.

Since then a succession of inventions have sought to keep the solo diver submerged at sea level pressure regardless of depth, thus obviating the need for decompression - ever.

Most of these ingenious devices were not suits as much as they were observation chambers lowered into the water on cables, and controlled by maneuvering the ship according to instructions telephoned from below. Some of the more famous models were the Davis Observation Chamber, Neufeldt and Kuhnke's Iron Duke, and William Beebe's Bathysphere. Observation chambers lacked two features essential to underwater work: mobility and manipulation.

Armored diving suits solved the latter problem. These were one-atmosphere chambers that were shaped in the form of their creator: with a form-fitted compartment for body and the extremities, with articulated joints for arms and legs, and with mechanical grips at the ends of the hands. They gained a great deal of publicity during salvage operations in the 1930's.

These suits were designed with ball joints made of overlapping steel plates similar in form and function to a lobster's tail. Robby the Robot of *Forbidden Planet* fame was probably modeled after them. Although workers in armored suits could handle tools and perform tasks that did not require great dexterity, they lacked lateral mobility because they were still tethered to their surface support vessels.

In the evolutionary process of one-atmosphere diving systems both mobility and manipulation have finally been obtained. Jim, Sam, Wasp, and Newt suits are all variations on a theme whose commercial and scientific applications are multitudinous. They move through the water by means of small propellers and, in a pinch, some of them can walk along the bottom.

However, like their predecessors, these units must be deployed and retrieved by crane, and only governments and large commercial diving companies can afford to rent or purchase them and the equipment needed to support them. In truth, these "suits" are not suits at all, but miniature one-person submarines that have forsaken the traditional torpedo shape and have assumed the guise of mortal man like an underwater golem.

The Generic Rebreather

So what can we do to stay deeper longer once we've outgrown the capacity of double tanks and a sled full of stage bottles, and can't afford the cost and constraints of armored dress? If you've read the chapter title then you know that I'm leading up to self-contained gas delivery systems called "rebreathers."

In general, a rebreather is an apparatus that recirculates some or all of a diver's exhaled breath and lets him inhale the unused oxygen again. In the recirculation process, the metabolic byproduct - carbon dioxide - is removed or "scrubbed."

Let's take the hype out of hypothetical and look at the practical side of rebreathers. The inside.

The major components of a modern generic rebreather are the gas cylinders, a flexible mixing bag called a counterlung, a carbon dioxide scrubber, a water absorbent, a battery, electrical circuits, and, in truly sophisticated models, a computer with digital readout displays. These sub-systems are all connected by a vast interlaced network of tubes, hoses, valves, wires, and electronic sensors that make the human circulatory system seem simple by comparison.

One gas cylinder contains oxygen, the other contains an inert gas or diluent, the purpose of which is to provide volume so the partial pressure of oxygen can be maintained. The diluent can be either nitrogen (for diving shallow) or helium (for diving deep). The composition of the resultant mix between oxygen and diluent is calculated on the maximum planned depth of the dive. This is so the blend can be breathed straight in open-circuit mode should the closed-circuit system fail.

As already noted, helium is superior to nitrogen because it is not narcotic. Of lesser import to most technical divers, at depths in the range of 800 fsw the density of nitrogen produces more breathing resistance than helium, which in turn increases CO_2 retention. Comparatively speaking, helium is like water while nitrogen is like molasses (treacle in the U.K.).

Exhaled gas goes through a water trap that removes excess moisture, through a CO_2 scrubber, and into the mixing bag. The purified gas is analyzed by an oxygen sensor. Additional oxygen is added to make up for the oxygen that has been consumed. An external pressure sensor determines the ambient pressure, which in turn triggers the release of diluent into the mixing bag as necessary. The resultant blend is ready to inhale at a precise, predetermined partial pressure of oxygen.

A side benefit of the scrubbing process is a thermal contribution. The chemical reaction generates heat which is imparted into the breathing loop and passed through the lungs, providing warmth to the core.

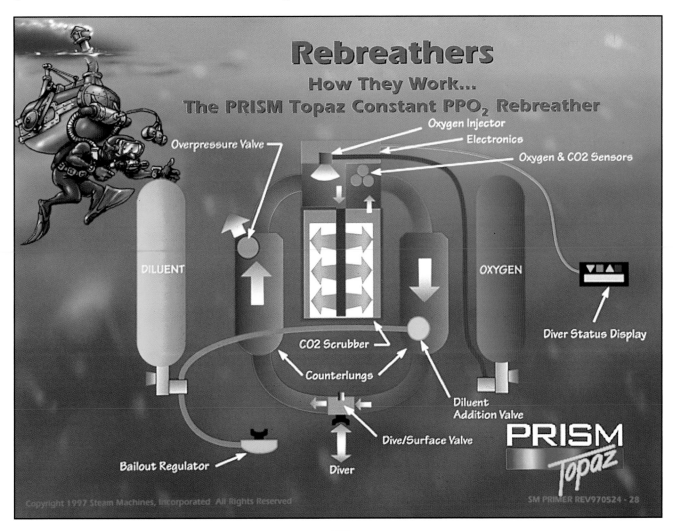

Our generic rebreather incorporates such safety features as non-return valves, manual bypasses, malfunction warning lights, and emergency bail-out capability (short-term open-circuit mode). No one wants bad breath in deep water.

The chief deterrents are purchase price, specialized training, maintenance costs, replacement parts such as sensors and scrubber material, and preparation of the unit and pre-dive systems check-out time. Prepping a rebreather for a dive is not like slapping a regulator onto a tank. The user needs extensive training in the unit's complex operational idiosyncrasies.

Field scientists testing rebreathers complain that they are frustrated by the amount of time spent on apparatus preparation and maintenance. This equates to time taken away from scientific research. Technical divers, who find the technology appealing for its own sake or who are driven by the thrill of exploration, are less likely to view set-up and care of equipment as a nuisance because working on and reconfiguring gear is an accepted prerequisite to underwater activity.

A not-so-obvious advantage of rebreathers is improved wireless communications due to the nearly silent operation: the only audible noise results from gas flow, clicking solenoids, the opening and closing of tiny valves, and the expansion and contraction of the mixing bag.

Recycling Oxygen

As with most other devices discussed in this volume, rebreathers have a history behind them, in this case more than a century old. The first rebreather hit the water in 1878. It relied on hemp impregnated with potash to remove the carbon dioxide from the diver's exhaled air. The potash was caustic and caused burns if it touched the skin or lips.

As already mentioned in the chapter on "Diver Propulsion Vehicles," Italian commandos in World War Two used oxygen rebreathers to great advantage in underwater military operations. The unit consisted of a full-face mask, a cylinder of oxygen, and a canister filled with soda lime (a carbon dioxide absorbent). They were taken down to depths of 50 fsw, although, after a several trainees convulsed and died, the survivors were warned to watch out for twitching of the extremities: a sure sign of oxygen toxicity. The duration of the unit was one to two hours.

The British navy was not far behind with an oxygen rebreather of its own. It had a duration of nine hours.

The U.S. Navy developed a unit that was officially designated as the Recirculating Underwater Breathing Apparatus, Closed-Circuit, Oxygen "for missions in which its bubble-free characteristics are essential for undetected approaches to objectives."

The military was interested primarily in submarine escape units and shallow water, clandestine combat missions such as storming enemy beaches and securing limpet mines to the hulls of enemy warships. Commercial diving outfits relied on surface-supplied air or scuba since the obvious exhaust of bubbles didn't matter.

For technical diving purposes, a rebreather with a depth limitation of 25 fsw isn't worth having. Long durations in shallow water can be more cheaply and easily achieved by means of ordinary scuba with large-capacity tanks, or with hookah: a unit consisting of a small, gasoline-powered compressor that delivers low-pressure air through a long hose. The compressor can be floated on an automobile inner tube in order to provide mobility.

For decades that is where the matter stood. There didn't seem to be a need for a rebreather that could be used in deep water, where pure oxygen became poisonous.

Mixed-gas Rebreathers

Eventually, the need to conduct untethered deep-water commercial assignments reached the point where a marriage between helium and rebreathers became desirable. Early examples in the U.S. were the Beckman Electro-Lung and the Biomarine CCR-1000.

The concept of mixed-gas rebreathers (MGR's) began to catch on in the military mind when the U.S. Navy realized that special warfare or explosive ordnance disposal might have to be carried out under conditions in which depth, duration, and clandestine maneuverability were valid considerations. Thus was born the concept of the EX 19 mixed-gas rebreather.

Years of painstaking research, development, and testing were required before a project of such dimension and technological ambition produced a prototype. Many of the sub-systems are well-kept secrets. One press report stated that the unit could be used "in excess of six hours at in excess of 300 feet." Note the double usage of "in excess."

A report released by the U.S. Navy Experimental Diving Unit, Panama City, Florida, in 1989, is more explicit by its very title: "EX 19 Performance Testing at 850 and 450 FSW." Keep in mind that the duration of most closed-circuit rebreathers is independent of depth because the gases (except for carbon dioxide) are recirculated. If a unit will last for six hours at 300 fsw, it will last for six hours at 850 fsw.

For some reason the EX 19 was shelved before the unit went into production. The Navy opted instead to produce the Mk 16, whose capabilities are classified.

An MGR in Civvies

Spurred by scientific applications such as underwater archaeology, marine biology, human physiology, and geological exploration, many nations have established mixed-gas rebreather research programs. Private companies have also entered the arena. Their primary goal is to adapt rebreathers that were originally built to military specifications (mil-spec) for non-military use in the scientific or recreational diving communities.

Because of strict design criteria, the military units are cost-prohibitive for anything other than combat missions. By contrast, a technical diver doesn't need equipment that is camouflaged, super silent, anti-magnetic, and bullet proof. (Unless he's fighting state and federal authorities over his right to locate and dive on shipwrecks.) One game that is now afoot is to scale down Navy MGR's and make appropriate adaptations in order to produce a unit that is cost-effective for widespread civilian use.

Taking a different tack, some manufacturers have undertaken independent development programs whose premise is to start from scratch and integrate the sub-systems with the recreational market in mind.

It must be understood that an MGR is inherently more complicated than an oxygen recycler because it must do so much more than simply scrub exhaled gas of carbon dioxide. It must maintain a constant partial pressure of oxygen regardless of ambient pressure, and it must do this by mixing the precise amount of diluent at exactly the right time.

This means that the diver's gas mix must be constantly monitored by sensitive analyzers so that just the right amounts of oxygen and inert gas are imparted for the depth at which each breath is taken. The optimal MGR will be a self-contained gas blending unit that is controlled by a microprocessor whose sensors monitor depth, partial pressures, and, ultimately, decompression commitment.

Semi-closed: Half the Job

Not all rebreathers are created equal. One sub-species is the semi-closed rebreather, so called because the unit recycles only a portion of the exhaled gas while exhausting the rest into the water. In this case semi-closed means partly open.

Compared to open-circuit scuba, semi-closed rebreathers produce fewer bubbles, weigh a bit less, use a lower quantity of gas, and cost five to ten times as much to purchase and maintain. The SCR is a mixed bag of tricks.

The system is simplistic and dependable because it utilizes a two-stage constant-flow regulator and does not rely on electronics for operation. Gas is inhaled from one bag and exhaled into another bag (called counterlungs), after which it is forced through the scrubbing canister and back to the breathing bag. Additional breathing gas is provided from a cylinder with each cycle. Excess gas that is generated by the constant flow and is not consumed is vented through a valve.

Before each dive, the unit is charged with gas in which the partial pressure of oxygen is pre-set for the planned maximum depth. The depth range is limited by the toxicity factor the same as in open-circuit scuba. Straying too deep increases the risk. Using air as a diluent offers a bail-out option in open-circuit mode in the event of a recycling malfunction.

Gas flows through the breathing loop at a constant rate. A sensor in the counterlung monitors volume and strives to keep it constant. As the volume drops, a tilt valve triggers the release of diluent, which is then injected mechanically into the breathing loop. The flow rate is *not* affected by work load, so that in situations of extreme activity the trickle of oxygen might not meet increased metabolic needs.

Product literature of one nitrox model emphasizes that constant buoyancy is provided by the counterlung because there is little or no change in the breathing loop - one bag is deflated while the other is inflated - so the tendency to ascend and descend with each breath is overcome; and that the operation of the valves, solenoids, and venting is quieter than the noise created by the stream of exhaust bubbles from open-circuit scuba. It also stresses that not as much gas is used - each cylinder holds less gas than a standard pony bottle - but fails to mention the cost of the scrubber material, which must be replaced after every dive and which can be a time-consuming process.

The duration of this model is forty minutes at 130 fsw: less than what the average person can get from an ordinary set of doubles. This mixed bag of slight advantages and disadvantages seems hardly worth the price tag. Why bother?

Another model offers instead of enriched air nitrox a breathing mix with a reduced fraction of oxygen, held constant at 18%. The unit features multiple gas cylinders and a duration of four hours: a definite plus. The downside is the high percentage of nitrogen, which increases the decompression penalty and equivalent narcotic depth. Because of the reduced oxygen content the gas can be breathed deeper than EANx, but only to the bounds of the breather's narcosis tolerance.

None of this goes to say that semi-closed nitrox rebreathers do not have their place in technical diving, only that they offer little more than stopgap features at a time when more effectual models are available. At the same time, these low-end units might satisfy people who are fascinated by the mystique of rebreather technology and who want to be the first on the block to own one. The retail price makes them attractive.

When reading product literature, beware of intentionally obfuscating language and evasive dis-information. For example, one advertisement claims that "the unit uses a passive gas control system keyed to respiratory minute volume," which I interpret to mean that it delivers gas on demand the same as a demand scuba regulator. Pompous grandiloquence does nothing more than obscure the issue.

On the question of endurance a brochure promoting one model claims that it depends upon "work level, the water temperature, the type and amount of gas supply, and the restrictions placed on a diver by decompression obligations and oxygen toxicity concerns. You can dive as deep and long as your training, gas mix, and support equip-

ment requirements allow." As intended, this says precisely nothing.

If a manufacturer can't or won't supply facts and figures about its product, perhaps that is because the true facts and figures are unfavorable.

Semi-closed with Helium

The next step in the refinement of semi-closed rebreathers for the technical diving market is the use of helium as a diluent, in order to increase the depth potential of the unit. The mixed-gas version works much like its nitrox counterpart. Pre-mixed heliox is circulated through the breathing loop, carbon dioxide is scrubbed from the exhaust gas, and oxygen is added as required.

The depth limitation is set by the percentage of oxygen in the mix. This is pre-selected in order not to exceed the desired partial pressure of oxygen at the planned maximum depth of the dive. Helium is non-narcotic - a wonderful advantage over nitrogen - but the decompression penalties prescribed by breathing helium for long durations at depth (say, beyond 150 fsw) can be severe.

The most complicated part of the unit is the oxygen injection system. Oxygen is injected into the breathing loop at a controlled rate that is based upon the mass flow of inspired gas. The mass flow rate can be controlled by one of several means: a venturi or metered orifice, a non-critical orifice and needle valve, or two orifices in line: one critical and one non-critical. When a diver inhales, the reduction of volume in the breathing loop is measured by a differential pressure gauge placed across the orifice in the tube, then the precise amount of volume is replaced from the gas supply.

This system may sound complicated because of the unfamiliar terminology, but in actuality it is elegantly simple - at least as simple as a demand scuba regulator. The moving parts are controlled by pressure reduction that results from inhalation.

The most critical function is maintaining a safe oxygen level. As noted above, the diver controls the maximum partial pressure of oxygen by not taking the unit deeper than the particular mix allows. Different mixes permit dives to various spreads of depth. Equally as important is preventing the partial pressure of oxygen from falling below the minimum threshold needed to prevent hypoxia (losing consciousness). This condition is controlled by increasing the injection rate of oxygen.

An increase in the injection rate correlates to an increase in consumption. (Remember that in a semi-closed system a portion of the exhaust gas is vented.) To a certain extent the minimum partial pressure of oxygen can be maintained by initially charging the gas bottle with a high percentage of oxygen - which then limits the depth range. Reducing the percentage of oxygen in the supply gas increases the depth range, but may fail to achieve the required partial pressure of oxygen needed at shallower depths.

In other words, although it seems counterintuitive, a unit that can be taken deep cannot be taken shallow. An alternate breathing apparatus might be required for initial descent and for the later stages of decompression. This see-saw effect of varying the injection rate represents the primary drawback of the semi-closed mixed-gas helium rebreather.

The key attributes are low weight and relative cost. The unit weighs less than the double tank rig it replaces. The unit costs more than a nitrox unit, but not as much more as open-circuit

Bill Stone surfaces after spending twenty-four hours under water breathing from the MK-1. Calculations of remaining consumables showed that he could have stayed another twenty-four hours. His biggest problem during the test dive was boredom. (Courtesy of Cis-Lunar, photo by Edward Sheffe.)

Bill Stone tests the MK-1 in Wakulla Spring. The small stream of bubbles burping from the bottom rear of the unit is rising from a loose gas connection that was later corrected. (Courtesy of Cis-Lunar, photo by Pete Scoones.)

helium costs more than open-circuit air. This is due to the expense of helium, much of which is recycled through the breathing loop.

It should be noted that since the percentage of oxygen in the mix is fixed, and cannot be altered during a dive, decompression profiles cannot be accelerated without staging or carrying auxiliary tanks of deco gas.

It has been estimated that one can either fly or make a repetitive dive only five hours after the initial dive on heliox. This is because of the rapid elimination of helium.

Closed-circuit in Wakulla Spring

One civilian-type MGR that was designed and built from the ground up was the Cis-Lunar MK-1, the prototype for the production model MK-2R. Bill Stone, experienced cave diver and Cis-Lunar founder, tested the MK-1 during a ten-week project in 1987 to explore and map one of Florida's most awesome artesian aquifers: Wakulla Spring.

During the project the MK-1 was depth tested to 150 ffw (feet of fresh water). Dives became progressively longer as the systems and controls were checked out. The culmination was an endurance run of twenty-four hours - and that was using only half of the unit's capability. Stone determined that with the MK-1 he could have conducted a 16-hour penetration that would still have followed the thirds rule: breathing one-third of the available gas before turning around, thus leaving one-third for the return trip and one-third for back-up. This was an estimated fifty-to-one increase in range over the open-circuit, multi-tank, sled-assisted explorations that were taking place simultaneously in the cave.

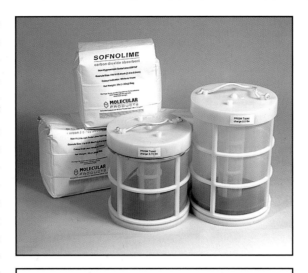

Prism Topaz radial flow scrubber baskets of varying durations. (Courtesy of Steam Machines, Inc., photo by Bo Mulder.)

Explore the Possibilities

Is there a place for this kind of breathing system in - dare I state it - "routine" technical diving?

You bet there is.

Even if you don't want to make hour-long dives to 300 fsw because of the decompression penalty, think about the incredible amount of reserve in case you had to extricate yourself from trouble such as entanglement or losing your way inside a cave or wreck. Why, in forty-eight hours you could even solve Rubic's Cube and still have time to feel your way out of the darkest, siltiest passageway, completely remove a trawler net from a wreck, take a compass bearing for shore, and decompress in the surf. (On second thought, I'm not sure about solving Rubic's Cube.)

But there's more.

The distinguishing feature of a closed-circuit mixed-gas rebreather is its ability to operate at a constant partial pressure of oxygen - unlike semi-closed systems, which operate at a constant *percentage* of oxygen with the partial pressure fluctuating as a function of depth. The ppO2 setting, called the set point, can be selected by the breather. Typically the partial pressure of oxygen is set at 1.4 ATA.

Once set, the MGR delivers gas to the breather at the optimum mix for the depth. As the breather descends and ascends, the partial pressure of oxygen is maintained by injecting discreet quantities of oxygen into the breathing loop: more or less as the metabolic process demands.

This exquisite quality goes far beyond keeping the breather safe from the harmful extremities of the oxygen partial pressure gradient and ensuring against toxicity and hypoxia. It optimizes the rates of both the absorption and elimination of inert gas by constantly minimizing the delivery of helium.

In the chapter on "Accelerated Decompression" I observed that decompression efficiency could be optimized by breathing a series of deco gases with increasingly elevated partial pressures of oxygen at each succeedingly shallower 10-fsw stop. The MGR does this automatically - and not just in 10-fsw increments but with each and every breath. The increments are graduated not just foot by foot - they are infinitesimally small.

This is such a transcendent advantage over open-circuit scuba that it can hardly be emphasized enough.

Prism Topaz in use. Note the bubble-free operation: silence that you can practically hear. (Courtesy of Steam Machines, Inc. photo by Bo Mulder.)

Examine the Realities

The more components a system contains, the more likely it is to fail. I am reminded of the wry observation made by one of the original seven astronauts as he waited to be launched into outer space, to the effect that the rocket beneath his seat was constructed of more than ten thousand moving parts, each one of which was supplied by the lowest bidder.

There is a tendency among people to distrust mechanical and electronic gadgets whose complexity is beyond their ability to understand. This goes extra for computers or any device that is computer driven because of glitches that may be minor in their manifestation but impossible for the average person to correct.

Without a doubt, the mixed-gas rebreather is a complex mechanism. It is synthesized from components that represent the latest technological advances in a wide range of disciplines. The integration of these various mechanical and electronic sub-systems requires the adherence to strict design criteria.

Yet one should not infer that there are inherent weaknesses in the system. A rebreather is far less complex than an automobile's computer controlled ignition system, internal combustion engine, and the integrated safety features such as steering controls and braking systems - all of which drivers take for granted.

The majority of rebreather accidents are caused by breather error: improper use, poor maintenance, inadequate preparation, failure to monitor warning signals, and general inattention to detail (such as forgetting to open the oxygen valve). The key to successful rebreather operation is unlearning ingrained open-circuit techniques and mastering closed-circuit rebreather skills.

Consider the Innovations

Beyond the generic rebreather are models of sophistication. Those who disclaim the reliability of computerized systems will be happy to note the graceful refinement which non-computerized rebreathers have embraced.

The Prism Topaz, designed and engineered by Peter Readey, is not computer controlled but relies strictly on analogue circuitry. The gas flow is mechanical. Galvanic sensors control the emission of oxygen into the breathing loop. Just as in an oxygen analyzer, the O_2 sensors use oxygen for

fuel and a battery to boost the signal for the display.

One flaw in early rebreather designs was flooding of the breathing loop if the mouthpiece

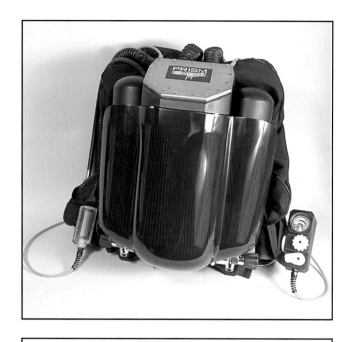

Prism Topaz shown from the back (above) and from the front (below). (Both courtesy of Steam Machines, Inc., photos by Bo Mulder.)

were dropped. Even if the hoses were purged, water reacted with the scrubber chemical and released caustic gas which seared the lips and lungs and made breathing impossible.

Nowadays, check-valves on either side of the mouthpiece prevent water from entering the hoses. No more "caustic cocktails."

Moisture from exhaled breath can cause blockages in the breathing loop tubing. Most of this is collected in the hose corrugations. Most of the rest is trapped by a water absorbent. Any left-over moisture has but little effect on carbon dioxide scrubbing material because present-day chemicals are coated to reduce reactivity.

CO_2 absorbents are more efficient and less expensive than older scrubbing materials. For example, Sofnolime is available in two grades: 4 to 8 fine mesh for normal use, 8 to 12 fine mesh for longer duration. The latter costs twice as much as the former, but even so the amount needed for each recharge is commensurate with the cost of an ordinary air fill. The coating also reduces abrasion and settling in the canister so the material doesn't turn to dust.

The location of the counterlung relative to the biological lungs can create a problem with hydrostatic pressure: the snorkel effect. In military units designed for mine disposal, the counterlung is backmounted because the breather ordinarily works in an upright position. But if the breather rolls over and looks up toward the surface, his cheeks may swell until he looks like a hamster hoarding food, and the mouthpiece may rip out his dentures if he isn't biting down hard. Over time, a hard continuous bite can cause jaw fatigue.

The Prism Topaz incorporates a split twin counterlung. The bags are mounted on the chest, one on either side and close to the mouth, similar in placement to the bladders in old-style BC's. This position is ideal for the common swimming attitude.

What about Back-up?

If you've read any other chapter in this book then you know that I'm a stickler for back-ups, especially where life support systems are concerned. Anti-progressives and rebreather detractors are quick to allege that MGR's contain unsatisfactory fail-safe mechanisms, that the failure of any single component or sub-system can cause the entire system to fail, and that total dependence on electronic monitoring can prove fatal in the event of water intrusion that shorts out delicate electrical circuits. But is this really true?

For all their high priced military specifications, Navy units do not incorporate back-ups. This is not to imply that the Navy is unmoved by safety concerns for its combat swimmers, or that casualties are expected in war by dint of the military mindset. On the contrary, it implies that the units are highly reliable and that Navy personnel are well trained and conversant with their use.

Consider again the Prism Topaz. It has a depth range of 500 fsw, a duration as long as six hours, an instrumentation panel that displays real-time data on all critical functions, an integral

Shown at left are the Steam Machines Model 1600 and the Prism Topaz, both with integrated buoyancy compensator and weight system. Shown at right is the Model 1600 with the cover removed so the interior is visible. (Both courtesy of Steam Machines, Inc, photos by Bo Mulder.)

buoyancy compensator - and all this in a 38-pound package!

The unit features an alternate solid scrubber cartridge that can tolerate flooding without emitting poisonous gas; not one but three separate oxygen sensors that operate in a "voting matrix" (they all deliver data simultaneously in order to reach a common consensus); and a secondary instrument display that is "used to verify primary display and control system and during manual operation with electronics offline."

Should all systems fail, the unit will deliver breathing gas in open-circuit mode via manual bypass valves.

Is this enough redundancy? All I can say is that diving is not a risk-free activity. Somewhere you've got to draw the line and accept the risks that are presented by the environment and by the equipment currently available. Where you draw that line is up to you.

Bill Stone's alternative to component redundancy is grandiose and expensive but satisfyingly practical: carry a complete spare rebreather. They're small enough and light enough. For extensive cave exploration he promotes back-mounting one rebreather and securing another to a diver propulsion vehicle. Furthermore, he advanced the idea of the primary unit being the one that is secured to the DPV, and keeping the backmounted unit as the back-up.

Don't Put Me On

Don't think you can just strap on a close-circuit underwater breathing apparatus (CCUBA) and head down into the submerged blue yonder. Special training is required before you can operate a rebreather safely. Nor is the purpose of this brief overview to teach anyone all the ins and outs of rebreathers. I merely want to dispel some of the misconceptions about them and to point out the unique features that set them apart from conventional scuba.

Will rebreathers eventually replace open-circuit scuba? Probably not. But they do offer options and present alternatives to conventional scuba for missions in which endurance at depth and for length of decompression cannot otherwise be achieved.

Rebreathers are costly and a lot of work to maintain. After every dive the breathing loop must be disinfected, the scrubber material must be replaced, and the gas bottles must be recharged. Before every dive the mechanical systems must be tested and the instrument displays verified. And in between trips new oxygen sensors must be installed.

On the other hand, whereas trimix is the gas of choice in open-circuit scuba, the closed-circuit technology of rebreathers is so efficient that the cost of helium - because it's recycled - is insignificant. Heliox is the gas of choice in closed-circuit rebreathers.

By integrating a decompression computer with the inert gas supply a custom decompression profile can be cut to order: an option that is not far off.

Is this a breath-taking concept? Not at all. It is reality that will give you breath where none was available before.

While oxygenated fluorocarbon liquid is a romantic idea that works only in laboratory rats and science fiction films, the mixed-gas rebreather actually exists - and at a price that is commensurate with all the tanks, gas, regulators, buoyancy compensators, and decompression computers that it replaces.

Now is the time to become a rebreather breather.

POST SCRIPT

If this book seems to have dwelt overlong on the practice of decompression it is because of the vitally important relationship between continued practice and survival. To comfortably explore the caves and wrecks that today's divers are discovering requires an exacting command and full understanding of practical applications. Then you can have a safe and memorable experience.

Technical dives must be planned with extraordinary forethought and executed with precision and control. Each diver must become a decompression specialist. Comprehend what this means.

It is not necessary for you to keep up with the latest surmises in decompression theory, or to have a thorough working knowledge of the biological processes that govern inert gas absorption and elimination. That is a job for scientists and doctors of hyperbaric medicine. Let them argue among themselves about what happens to the body during the various phases of pressure change.

Their specialty is study, yours is performance. They work in the clean air environment of laboratories, while you dare the airless medium under water.

The plain fact of the matter is that you can't be an expert in everything, nor do you need a Ph.D. in order to become an accomplished deep decompression diver. It is more important for you to know your physical limitations than to understand the mechanisms of bubble formation, of greater consequence to recognize the symptoms of oxygen toxicity than to determine its underlying causes. Which is not to say that an understanding of bubble formation and its deleterious effects won't help you appreciate the cruciality of perfecting decompression technique.

A hyperbaric physiologist once told me that I couldn't make the deep repetitive dives that I was routinely making because the Navy Tables said so. Like the bumble bees that continued to fly after physicists "proved" mathematically that their wingspan could not support their body weight in flight, I said, "Gee, no kidding," and went right on diving.

He thought I was crazy. I thought he was shortsighted and narrow-minded. He preferred the solace of his beliefs to the empirical evidence before his eyes. Today, more people recognize that almost anything can be done if you are willing to make the necessary sacrifices. Dedication overrules adversity.

In the world of technical diving, the way to achievement is through training. You will never become an expert by reading books - not even this book. Perhaps you don't *need* course instruction in established diving procedures, but why re-invent the wheel? Sure you can learn it all on your own. I did. But it took two decades to master what today's technical entrants can be taught in a couple of years.

Instruction and course work notwithstanding, you *do* need to observe closely how others have perfected their craft, and you need to try unfamiliar techniques in order to determine which ones will work for you. By keeping an open mind you can then adapt those techniques to suit the circumstances you encounter, or you can develop techniques of your own.

Then, after absorbing all you can, get into the water where the conditions are variable and less than ideal. You must become a practitioner and constantly strive to transcend the teachings of those who came before you.

Throughout this book I have endeavored to demonstrate that diving techniques and equipment development and configurations have not reached the last stages of perfection. Nothing ever does.

Free-thinking individuals will always strive to create new concepts, experiment with new methods, and seek more lofty goals.

That is human nature. Technical diving has arrived at a turning point in an exponential curve where growth is accelerating with ever-quickening speed. A brand new generation of underwater explorers is being born.

Technical diving is a progressive adventure of which everyone can be a part. Ahead are new frontiers that need pioneers to lead the way. Stay with it. Vaulting challenges and stark excitement are waiting over the next horizon.

Savor the thought of meeting them.

Pete Manchee and Ed Suarez decompress after a dive to the *Ostfriesland.*

GLOSSARY
Acronyms, Initialisms, Buzz Words, and Technical Terms

apoxia: total lack of oxygen reaching body tissues, which can lead to death. (Compare hypoxia.)

Ar: the chemical symbol for argon.

ATA (or ata): atmospheres absolute; the pressure at sea level equivalent to 14.7 psi.

bar: a metric unit of pressure equal to 14.5 psi, or slightly less than one atmosphere (which is 14.7 psi).

BC (or BCD): buoyancy compensator (or buoyancy control device); an inflatable bag worn to control buoyancy and to compensate for the compression of neoprene and air at increased pressure.

BMF: backmounted flotation; see "wings."

bottom mix: a gas breathed on the bottom; generally a helium blend although air or nitrox can be bottom mixes.

Buhlmann, A.A.: the designer of high-altitude decompression tables used in some of today's decompression computers. Buhlmann's model is more accurate and more conservative than the old Haldanian model.

CCUBA: closed circuit underwater breathing aparatus; a type of rebreather.

cfg: cubic feet of gas.

CGA: Compressed Gas Association.

CNS: central nervous system.

comm unit: communications unit.

component pressure: the pressure in pounds per square inch of a single constituent gas in a breathing mix; for example, the component pressure of oxygen in a tank of nitrox-32 filled to 3,500 psi is 487 psi.

composite tank: a lightweight scuba cylinder consisting of an aluminum liner covered with Kevlar or some similar spun hydrocarbon.

converter block: the high-pressure fitting at the in-water end of the hose in a surface-supplied oxygen rig, and that is the mating surface for the regulators; it is available in either screw-and-yoke or DIN.

DCAP: Decompression Computation and Analysis Program, a proprietary computer program designed by Bill Hamilton to generate decompression schedules for a variety of breathing mixes; pronounced "dee'-cap."

DCI: decompression injury or illness, the acronym that is replacing DCS.

DCIEM: Defence and Civil Institute of Environmental Medicine, the designer of decompression tables for the Canadian navy.

DCS: decompression sickness, also know as caisson disease and the bends; now being replaced by the acronym DCI.

deco: short for decompression

deco bottle: a cylinder that is used to carry gas for decompression.

deco gas: a gas that is used for decompression purposes, especially accelerated decompression. It generally refers to oxygen or to a nitrox blend, but it could mean a helium blend with a lower percentage of helium than the bottom mix.

decompression schedule: any single decompression profile that is computer generated for a specific gas mix (including air) for a specific depth and duration, differentiated from decompression table.

decompression table: a printed version of cumulative decompression data which includes a wide range of depths and durations for a specific gas mix (including air), and formatted in rows and columns for ease of comparative views at a glance. Examples are the United States Navy Standard Air Decompression Tables and the DCIEM tables. Differentiated from decompression schedule.

diatomic: consisting of two atoms with shared electrons.

diluent: any inert gas that provides volume to a mix in order to keep the partial pressure of oxygen within prescribed limits. Nitrogen is a diluent in air; helium is used as a diluent in mixed gas.

DIN: Deutsche Industrie Norme (German Industry Standard); fittings made to these specifications employ a captured o-ring assembly for use in high-pressure applications.

Doppler: see "ultrasonic Doppler."

Doppler tables: decompression tables whose parameters were tested by means of an ultrasonic Doppler; also called "no bubble" tables because the time limits are such that no perceptible bubbles of inert gas are allowed to form in the blood stream.

DOT: Department of Transportation; a U.S. regulatory authority with more than thirty agencies under its control, concerned primarily with national transportation and environmental safety. Among its many responsibilities is the regulation of scuba cylinder specifications. The Coast Guard is a DOT agency.

DPV: diver propulsion vehicle; a waterproof housing for a battery-operated motor which turns an external propeller in order to pull a diver through the water. The small varieties are generally referred to as "scooters."

EAD: equivalent air depth. For decompression computation purposes, the correlation of the partial pressure of nitrogen in a particular nitrox mix compared to the partial pressure of nitrogen in air. For example, the depth of 130 feet on nitrox-32 is approximately equivalent to the depth of 110 feet on air.

EANx: enriched air nitrox; any nitrox blend that contains a higher partial pressure of oxygen than that contained in ordinary air.

END: equivalent narcotic depth. The degree of narcosis felt from the reduced partial pressure of nitrogen in a mixed gas at depth compared to the narcosis felt on air at a shallower depth. For example, breathing trimix-19/30 at 230 fsw yields the same degree of narcosis as breathing air at 140 fsw.

ffw: feet of fresh water.

fHe: fraction of helium.

fN$_2$: fraction of nitrogen.

fO$_2$: fraction of oxygen.

fsw: feet of sea water.

Haldane, J.S.: the early 1900's designer of decompression tables whose half-time model has been the basis for most of the decompression tables generated during the twentieth century; the Haldanian model has been modified many times since, and has now been largely replaced by more accurate and more conservative models.

He: the chemical symbol for helium.

heliox: a blend of oxygen and helium. There is no nitrogen in heliox.

Huggins, Karl: the designer of decompression tables used in some of today's decompression computers.

hydraliox: a breathing gas consisting of a mixture of hydrogen, helium, and oxygen. It has been breathed by divers down to depths of 1,700 feet.

hydrox: a breathing gas consisting of a mixture of hydrogen and oxygen.

hypoxia: deficiency in the amount of oxygen reaching bodily tissues; it can lead to loss of consciousness. (Compare apoxia.)

isobaric counterdiffusion: a little understood chemical and physiological mechanism in which one gas diffuses through a cell wall in one direction without interrupting the diffusion of another gas in the opposite direction, if the gases have different diffusion rates. In practical terms, this means that one gas is absorbed while another is eliminated, neither one blocking the progress of the other, and in some cases one gas accelerating the progress of the other.

narcosis: the narcotic effect produced by an inert gas in a breathing mix; for example, nitrogen narcosis is the narcotic effect of nitrogen in air under pressure.

MGR: mixed-gas rebreather; see rebreather.

mix (or mixed gas): any combination of gases other than atmospheric air; it generally refers to mixes other than nitrox, and in technical diving vernacular usually means blends of helium such as heliox and trimix.

mph: miles per hour.

msw: meters of sea water.

N_2: the chemical symbol for nitrogen; the "2" refers to nitrogen's diatomic structure.

neox: a breathing gas consisting of a mixture of neon and oxygen.

nitrox: a breathing gas consisting of any proportion of nitrogen and oxygen, including air, but in technical diving vernacular generally referring to mixtures in which the percentage of oxygen is higher than that of air; in this context it is more accurately called "oxygen enriched air" or "enriched air nitrox."

NOAA: National Oceanic and Atmospheric Administration; pronounced as "Noah."

NOAA nitrox I (or NNI): a standardized nitrox blend consisting of 32% oxygen and 68% nitrogen; due to the potential for incurring oxygen toxicity, its use is limited to a depth of 130 fsw.

NOAA nitrox II (or NNII): a standardized nitrox blend consisting of 36% oxygen and 64% nitrogen; due to the potential for incurring oxygen toxicity, its use is limited to a depth of 110 fsw.

normoxic: a mixture of nitrogen and oxygen used in underwater habitats; the ratio of mix is selected to provide a partial pressure of oxygen that is close to or slightly above the normal atmospheric value of .21 ATA at the saturated depth.

O_2: the chemical symbol for oxygen; the "2" refers to oxygen's diatomic structure.

OTU: oxygen toxicity unit; a measure of exposure to oxygen in a breathing medium. Units are accumulated according to durations spent at all pressures. The sum of accumulated units represents a person's total exposure, from which the potential for incurring pulmonary of whole-body oxygen toxicity may be inferred (but not CNS oxygen toxicity).

oxtox: oxygen toxicity.

oxygen toxicity: physiological damage produced by breathing high partial pressures of oxygen. Initial symptoms include muscular twitching of the face and lips, tingling sensations in the fingertips, nausea, dizziness, tunnel vision, and ringing in the ears. Advanced symptoms include fatigue, confusion, anxiety, and uncoordinated or random movements. The final stage is convulsion and unconsciousness. The final stage may occur instantaneously without the warning of other symptoms.

partial pressure: that portion of the total pressure of a breathing medium exerted by a single gas in the mix. This value is variable because it is affected by changes in ambient pressure. For example, the partial pressure of oxygen in air at sea level is .2095 atmospheres absolute, whereas at 297 fsw it is 2.095 atmospheres absolute; differentiated from percentage.

percentage: the fraction of a particular gas in a mix compared to the total. This value is not affected by changes in ambient pressure. For example, the percentage of oxygen in air is 20.95% regardless of the amount of pressure exerted on the air; differentiated from partial pressure.

ppHe (or pHe): partial pressure of helium.

ppN_2 (or pN_2): partial pressure of nitrogen.

ppO_2 (or pO_2): partial pressure of oxygen. (Avoid writing PO_2 because it looks like the chemical symbol for phosphorus oxide, and may be confusing to some.)

psi: pounds per square inch.

PVC: polyvinyl chloride (a polymerized thermoplastic resin).

rebreather: a self-contained breathing apparatus which recycles some or all of the exhaled gas after passing it through a scrubber that removes the carbon dioxide. Open-circuit rebreathers exhaust some of the exhaled oxygen, closed-circuit rebreathers recycle all the exhaled oxygen. Oxygen rebreathers contain only oxygen; mixed-gas rebreathers contain oxygen and a diluent such as helium or nitrogen (or a combination of both).

reverse gas diffusion: a condition in which an inert gas going into solution in the tissues prevents elimination of the same gas from a previous dive or from a denser inert gas contained in the breathing mix on the same dive, thus altering the actual decompression profile calculated by the schedule produced for the particular mix. For example, in trimix the uptake of nitrogen slows the offgassing of helium, a fact that must be taken into consideration when designing a decompression program.

RNPL: Royal Navy Physiological Laboratory, the designer of decompression tables for the British navy.

ROV: an unmanned remotely operated vehicle "flown" from a surface support vessel by means of cables, using integral video cameras to let the operator see its movements; pronounced "rove" or as three individual letters.

SAC rate: surface air consumption rate.

SPG: submersible pressure gauge.

stage bottle: a scuba cylinder carried in addition to those tanks intended for primary use, and placed or staged at a specified location during descent or during the outbound phase of a dive, for later retrieval; a stage bottle may contain gas for decompression, or may be used to extend the range of exploration.

surface decompression: decompression conducted in a pressurized chamber on the deck of a boat; called "Sur-D" for short.

Sur-D O2: surface decompression while breathing oxygen; this is conducted inside a decompression chamber, not on the ground or the deck of a boat.

thirds rule: the standard cave-diving practice in which one-third of the total available breathing gas is used during the outbound phase of the dive and one-third is used during the return, leaving one-third for a safety margin and for emergency use.

trimix: a blend of oxygen, helium, and nitrogen.

ultrasonic Doppler: a high-frequency (ultrasound) flowmeter whose sensor is placed on strategic parts of the body so the device can detect bubbles circulating in the bloodstream; it is called a flowmeter because it detects only moving bubbles, not those trapped in the tissues.

VIP: visual inspection performed; a scuba cylinder is inspected annually for signs of rust, flaking, or water.

VOX (or vox): specifically, voice activated transmission; also, a generic acronym used to describe any kind of wireless underwater communation device.

whip: a hose; it can refer to an ordinary regulator hose, but most often is intended to mean a hose of longer length and used for a different purpose; for example, an oxygen whip may be a hose (either high-pressure or low-pressure) that is hung over the side of a boat for decompression purposes; a fill whip is a hose used to transfer gas into a cylinder; a drysuit whip is an inflator hose; and so on.

wings: a backmounted buoyancy compensator: one that is worn between the tanks and the backplate; so called because the inflatable bag stretches to the sides like a pair of wings.

INDEX

Books by the Author

Fiction

Vietnam
Lonely Conflict

Action/Adventure
Memory Lane
Mind Set

Supernatural
The Lurking

Science Fiction
Entropy
Return to Mars
Silent Autumn
The Time Dragons Trilogy
A Time for Dragons
Dragons Past
No Future for Dragons

Nonfiction

Advanced Wreck Diving Guide
Ultimate Wreck Diving Guide

Track of the Gray Wolf
Shipwrecks of New Jersey

Available (postage paid) from: GARY GENTILE PRODUCTIONS
                                    P.O. Box 57137
                                    Philadelphia, PA 19111

Nonfiction

$25   *Andrea Doria: Dive to an Era* (hard cover)
$20   *The Nautical Cyclopedia*
$20   *USS San Diego: the Last Armored Cruiser*
$20   *Wreck Diving Adventures*
$20   *Primary Wreck Diving Guide*
$30   *The Technical Diving Handbook* (large format)
       Civil War ironclad MONITOR
$25   Book (hard cover) *Ironclad Legacy: Battles of the USS Monitor*
$25   Videotape (VHS or PAL): *The Battle for the USS Monitor*
       The Popular Dive Guide Series
$20   *Shipwrecks of New York*
$20   *Shipwrecks of Delaware and Maryland*
$20   *Shipwrecks of Virginia*
$20   *Shipwrecks of North Carolina:*
           from the Diamond Shoals North
$20   *Shipwrecks of North Carolina:*
           from Hatteras Inlet South
Wreck Diving Adventure Novel
$20   *The Peking Papers* (hard cover)

Website - http://www.pilot.infi.net/~boring/gentile.html